Pathways and Challenges for Efficient Desalination

Edited by Muhammad Wakil Shahzad, Mike Dixon, Giancarlo Barassi, Ben Bin Xu and Yinzhu Jiang

Published in London, United Kingdom

IntechOpen

Supporting open minds since 2005

Pathways and Challenges for Efficient Desalination
http://dx.doi.org/10.5772/intechopen.94651
Edited by Muhammad Wakil Shahzad, Mike Dixon, Giancarlo Barassi, Ben Bin Xu and Yinzhu Jiang

Contributors
Thallam Lakshmi Prasad, Nicholas Nelson, Antonella De Luca, Khaled Boughzala, Mustapha Hidouri, Saeed Pourkarim Nozhdehi, Issam Daghari, Ishita Shrivastava, Edward Eric Adams

First published in London, United Kingdom, 2022 by IntechOpen
IntechOpen is the global imprint of INTECHOPEN LIMITED, registered in England and Wales, registration number: 11086078, 5 Princes Gate Court, London, SW7 2QJ, United Kingdom
Printed in Croatia

British Library Cataloguing-in-Publication Data
A catalogue record for this book is available from the British Library

Additional hard and PDF copies can be obtained from orders@intechopen.com

Pathways and Challenges for Efficient Desalination
Edited by Muhammad Wakil Shahzad, Mike Dixon, Giancarlo Barassi, Ben Bin Xu and Yinzhu Jiang
p. cm.
Print ISBN 978-1-83968-876-8
Online ISBN 978-1-83968-877-5
eBook (PDF) ISBN 978-1-83968-878-2

For EU product safety concerns:
IN TECH d.o.o., Prolaz Marije Krucifikse Kozulić 3, 51000 Rijeka, Croatia, info@intechopen.com or visit our website at intechopen.com.

Meet the editors

Dr. Muhammad Wakil Shahzad is a senior lecturer in the Mechanical and Construction Engineering Department, Northumbria University (NU), Newcastle Upon Tyne, United Kingdom. He is an expert in hybrid desalination processes and renewable energy applications for water treatment. He has won many international awards for his innovative desalination cycle, including the National Energy Globe Award Saudi Arabia 2021, Sustainability Medal 2020, Global Innovation Award 2020, National Energy Globe Award Saudi Arabia 2020 and 2019, Excellence and Leadership Award 2019, and IDA Environmental & Sustainability Award 2019. His research has been highlighted at Yahoo Business, Nature Middle East, Arab News, and many other national and international platforms. Dr. Shahzad has a Ph.D. in Mechanical Engineering from the National University of Singapore and research training from KAUST Saudi Arabia. He has extensive experience in intellectual property development and the commercialization of innovative technologies. He holds eleven international patents. To date, he has published two books, seventeen book chapters, more than seventy peer-reviewed journal papers, and more than 110 conference papers. He also received three best paper awards in international conferences. He is an editor of *International Communications in Heat and Mass Transfer*, an editorial board member of SN Applied Sciences, and a guest editor for topical collections. He is a Chartered Engineer and a mentor for International Desalination Association's Young Leader Program (IDA-YLP). He is also a member of many professional organizations, including the Institute of Mechanical Engineers, International Desalination Association (IDA), International Water Association (IWA), and American Society of Mechanical Engineers (ASME).

Dr. Mike Dixon is a desalination and water treatment technology professional with 15 years' experience working with membrane and thermal technologies in Australia, North America, the Middle East, the Caribbean, and Asia. He has worked across the entire value chain with technology manufacturers, water utilities, oil and gas companies, pharmaceutical companies, and research hubs. As the current Chief Technology Officer of WaterNEXT (a water business accelerator based in Calgary, Canada), Dr. Dixon's responsibilities include technology assessment and intake assessment. He is also a senior technical advisor, and on the founding team, of MIT start-up Sandymount Technologies. Prior to WaterSMART, Dr. Dixon was Applications Development Manager for NanoH2O, a global provider of reverse osmosis membranes that leveraged UCLA developed nanotechnology developed at the University of California, Los Angeles, to lower the cost of desalination with more than 300 installations in 40+ countries in the three years from market launch. LG Chem acquired NanoH2O in 2014. He was also R&D Engineer for the 300,000 m3/day Adelaide Desalination Plant. Dr. Dixon has a Ph.D. in Chemical Engineering from the University of Adelaide, Australia, and business training from Stanford University, California. He is experienced with the development of intellectual property and the commercialization of new technologies. He has more than sixty publications in international journals and is an author of several books and book chapters. He has been an editor of the *International Desalination Association Journal* and a reviewer for the *Journal of Membrane Science, Desalination and Water Research*.

He was National President of the Young Water Professionals for the Australian Water Association and in 2012 won the prestigious International Desalination Association Fellowship Award.

After finishing his Ph.D. in Chemistry in New Zealand, Dr. Giancarlo Barassi returned to his home country Chile and started his own company H2OPRO where he got involved in selling projects and equipment for Culligan, Voltea and FEDCO in Chile and Peru. He has hands-on experience in design, commissioning and troubleshooting of RO plants for Industrial, Agriculture and Haemodialysis applications. Dr. Barassi joined the pump and ERD manufacturer, FEDCO, in September 2017 where he collaborated in Sales and Business Development in the Western Hemisphere until 2021. He was appointed the co-chair of the Young Leader Program at the International Desalination Association for the 2019-2022 term where he promotes benefits to young professionals in our industry. In 2022 he joined Aquatech International LLC, as the Desalination and Reuse Market Manager, where he oversees all Desalination and Reuse projects.

Dr. Ben Bin Xu is a Full Professor of Materials and Mechanics, Department of Mechanical and Construction Engineering (MCE), Northumbria University (NU), Newcastle Upon Tyne, United Kingdom. and leads the research group of the university's Smart Materials and Surfaces Lab. He obtained his Ph.D. in Mechanical Engineering at Heriot-Watt University, Scotland, in 2011 and subsequently worked as a postdoc researcher at the University of Massachusetts Amherst (2011–2013). His research interests include advanced materials, smart surfaces, energy materials/systems, thin films, flexible electronics, and micro-engineering. He has published more than 120 journal articles and 3 book chapters. He has four patents to his credit. He has given more than sixty invited talks and won multiple awards, such as the Outstanding Paper Award at the 2018 IEEE International Flexible Electronics Technology Conference, 2016 Young Investigator Award from the International Polymer Networks Group, and 3rd prize in the 2016 EPSRC photo competition, among others. Dr. Xu is a fellow of the Royal Society of Chemistry (RSC), Institute of Materials, Minerals and Mining (IoM3), and Royal Society for the Encouragement of Arts, Manufactures and Commerce (RSA). He is also a Chartered Scientist and Chartered Engineer. He has been an elected member in the Advisory Council and Nomination Committee of IoM3 since 2019, an elected member in the RSC Materials Chemistry Division Council since 2020, an elected member in the RSC Newcastle upon Tyne and North East Coast Committee since 2020, and the executive committee for the Chinese Society of Chemical Science and Technology (CSCST-UK) since 2016.

Dr. Yinzhu Jiang is a professor at the School of Materials Science and Engineering, Zhejiang University, China. He received his Ph.D. in Materials Science and Engineering from the University of Science and Technology, China, in 2007. He worked as a post-doctoral researcher at Heriot-Watt University, Scotland, from 2007 to 2008, and an Alexander von Humboldt Fellow at Bielefeld University, Germany, from 2008 to 2010. His research interests focus mainly on energy-related materials and electrochemistry, including rechargeable batteries, metal anodes, and solid electrolytes. He has published more than 100 articles in journals such as *Advanced Materials, Advanced Energy Materials, Energy & Environmental Science, Nano Energy*, and others.

Contents

Preface

Water and energy are closely interlinked and interdependent valuable resources that underpin economic growth and human prosperity. In every part of daily life, such as power generation, feedstock crop production, and fossil fuel processing, water is a ubiquitous source. Similarly, energy is vital for powering the water cycle, which includes collection, treatment, and distribution to end users. The mutual vulnerability of water and energy is amplifying due to rising demand because of exponential growth of gross domestic product (GDP), increasing population, and climate change.

Global water demand is projected to increase more than 55% by 2050 mainly due to a high GDP growth rate that will increase water demand for manufacturing, power generation, and domestic sector use by 400%, 140%, and 130%, respectively. This current demand trend will place 40% of the world's population at risk of water scarcity by 2050. Presently, more than 19,500 desalination plants in 150 countries produce roughly 38 billion m^3 of water per year. This number is projected to increase to 54 billion m^3 per year by 2030, which is 40% more compared to 2016. Desalination is the most energy-intensive water treatment process, consuming 75.2 TWh or about 0.4% of global electricity per year. **Figure 1** shows the desalination capacities of the world and Gulf Cooperation Council (GCC) countries and their share of different technologies.

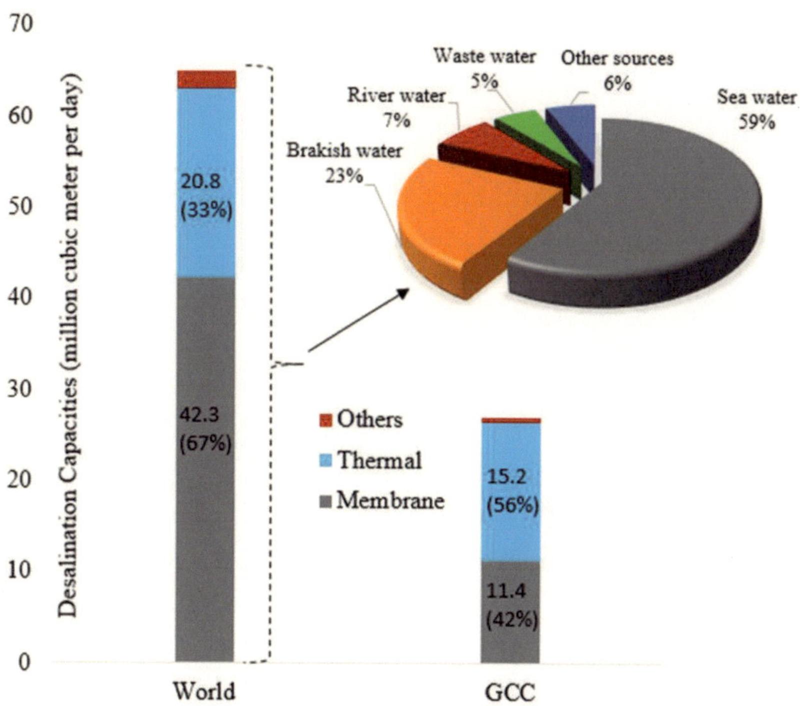

Figure 1.
World desalination capacities [1].

The conventional desalination technologies can be divided into two main categories: membrane separation (reverse osmosis [RO]) and thermally driven processes (multi-stage flash distillation [MSF], multi-effect distillation [MED], and adsorption desalination [AD]). In addition, forward osmosis (FO), capacitance deionization (CDI), membrane distillation (MD), and freezing and humidification dehumidification (HDH) are still in the research and development stages. Hybrid technologies such as MED–AD, MSF–MED, and RO–MSF have potential to overcome the limitations of conventional processes for greater recovery and performance. Unfortunately, conventional desalination processes are energy-intensive and non-eco-friendly. They only operate at 10–13% of their thermodynamic limit, which is not sustainable. To achieve UN sustainability goals, they should operate at more than 30% of the thermodynamic limit [2–6].

This book addresses key challenges related to the desalination industry. Energy recovery is an important parameter for efficient operation of a desalination plant. Chapter 1 provides details of energy recovery of membrane processes. Chapter 2 discusses membrane management. In an RO plant, the membrane is usually replaced every 3–5 years due to blockage and fouling. Membrane management is very important for reliable operation. Typically, recovery varies in the range of 40–45% of all desalination processes. Chapter 3 provides an outline for brine management to reduce marine pollution. Chapter 4 highlights water utilization in the agricultural sector. Typically, 70% of water is used for agriculture due to poor practices. This chapter presents best practices and efficient water utilization strategies. Chapter 5 provides an overview of the remineralization process after desalination. Remineralization is very important for achieving the World Health Organization's standards for drinking water quality. Chapter 6 discusses pre-treatment chemicals and how their injection can be optimized. Optimization of chemical injection in pre-treatment can lead to sustainable desalination and reduction of marine pollution.

Dr. Muhammad Wakil Shahzad
Department of Mechanical and Construction Engineering,
Northumbria University,
Newcastle upon Tyne, United Kingdom

Mike Dixon
Synauta Inc.,
Calgary, Canada

Giancarlo Barassi
Aquatech International LLC,
United States of America

Ben Bin Xu
Northumbria University,
Newcastle upon Tyne, United Kingdom

Dr. Yinzhu Jiang
Zhejiang University,
Hangzhou, China

References

[1] Shahzad MW, Burhan M, Ang L, Choon Ng K. Energy-water-environment nexus underpinning future desalination sustainability. Desalination 2017;413:52–64. https://doi.org/10.1016/j.desal.2017.03.009.

[2] Shahzad MW, Burhan M, Ybyraiymkul D, Ng KC. Desalination processes' efficiency and future roadmap. Entropy 2019;21.

[3] Ng KC, Burhan M, Chen Q, Ybyraiykul D, Akhtar FH, Kumja M, et al. A thermodynamic platform for evaluating the energy efficiency of combined power generation and desalination plants. Npj Clean Water 2021;4:1–10. https://doi.org/10.1038/s41545-021-00114-5.

[4] Shahzad MW, Burhan M, Ng KC. A standard primary energy approach for comparing desalination processes. Npj Clean Water 2019;2:1–7. https://doi.org/10.1038/s41545-018-0028-4.

[5] Shahzad MW, Burhan M, Ng KC. Pushing desalination recovery to the maximum limit: Membrane and thermal processes integration. Desalination 2017;416. https://doi.org/10.1016/j.desal.2017.04.024.

[6] Ng KC, Thu K, Oh SJ, Ang L, Shahzad MW, Ismail AB. Recent developments in thermally-driven seawater desalination: Energy efficiency improvement by hybridization of the MED and AD cycles. Desalination 2015;356. https://doi.org/10.1016/j.desal.2014.10.025.

Section 1

Energy for Water Treatment

Chapter 1

Energy Recovery in Membrane Process

Saeed Pourkarim Nozhdehi

Abstract

One way in order to reduction energy consumption and providing the required water in both well-established technologies such as reverse osmosis (RO) and electrodialysis is use of the strengths of two or more processes through hybridization. Other key objectives of hybridization include increasing the capacity of the plant flexibility in operation and meeting the specific requirements for water quality. At this section, has been provided a critical review of hybrid desalination systems, and methods used to optimize such systems with respect to these objectives. For instance, coupling two process like as electrodialysis with RO is very effective in order to overcome the low recovery in RO systems. On the other hand, we can use for two or more processes such as RO with membrane distillation (MD) or zero liquid discharge (ZLD) for treatment of hypersaline feed solutions. At this section, also have been reviewed the applicability of salinity gradient power technologies with desalination systems and we identified the gaps that for effective upscaling and execution and implementation of such hybrid systems need to be addressed.

Keywords: energy recovery, desalination, hybrid systems, reverse osmosis, membrane

1. Introduction

Sustainable energy is the key solution for addressing major concerns about the future such as climate change, environmental protection, and balanced growth of the economy and society. In many nations at past two decades have witnessed advancement in economic development. However, industrial advancement, deterioration of the environment, energy shortage, the rapid economic growth and increasing demands of growing populations pose a huge threat for future generations [1–3]. For many years, economic development has been the key focus of many policy makers in sustainable development until the inception of the Kyoto protocol agreement in 1997, which includes environmental quality as a crucial variable for sustainable development [3]. According to global energy consumption, expected that electricity demands to be double in the next twenty-five years, so, major opportunities for innovation in energy production, storage, transmission and use of it have begun to open up. In particular, in order to improving the efficiency of the processes and reducing the global carbon footprint, there is a huge interest in sustainable energy technologies [3, 4].

Development of an approach to sustainable energy that addresses greenhouse gas emission, environmental concerns, availability of resources, social impact and cost is an immense challenge. The key focus for obtaining energy sustainability is the generation of energy with renewable energy sources and replace them slowly with power fossil fuels [5]. There is much research that has worked for developing the membrane sector, which emphasizes the use of renewable energy in membrane technology. Although the efficiency of the process is still a high priority. Recently, membrane technologies, especially, in the water and energy sector, have begun to play a basic role in developing the infrastructure for sustainable energy. Some of the membrane-based approaches that are currently adapted at an industrial scale include desalination by RO, membrane-based bioreactors (MBR) for pure water generation, lithium-ion batteries, and membrane-based fuel cells and CO_2 capture [6–8]. Many advantages of membrane technologies like flexibility, feasibility and adaptability have been able to decrease many concerns related to water scarcity and energy demands in recent years. However, with achievement to advancements in membrane-based technologies. we still need to improve affordability and costs.

2. Membrane technology and sustainable water generation

In the past decades, following the increase in freshwater demand, various techniques including multiple-effect distillation (MED), vacuum distillation, multi-stage flash distillation (MSF), and other membrane-based technologies, such as reverse membrane distillation (MD), osmosis (RO) and etc., in order to sea water desalination, have been developed. Among these technologies, some of the membrane-based techniques such as RO, MD and forward osmosis (FO), because of some advantages like as lower maintenance and operating costs, lower capital requirements and low energy consumption, are considered as suitable alternatives [3, 9].

2.1 Desalination

Desalination is a process which use for producing freshwater from either sea or brackish water, by removing the salt content either by membrane technologies or by a thermal distillation process.

As can be seen from **Table 1**. the membrane technologies, specifically the RO, mainly, because of lower energy requirements, are preferred over the other technologies. In different technologies, the specific energy consumption (SEC) varies widely and depending on the operation and process control as well as the quality of the produced water, this value might have further differed significantly for a particular technology.

2.2 Reverse osmosis (RO)

To date, for desalination and stress reduction due to depletion of available water resources, reverse osmosis (RO) is the key technology [1]. In desalination plant such as RO, membrane played a key role which is largely determine the separation performance of the overall plant (**Figure 1**). In several recent studied suggests that in ultra-permeable membranes (UPMs) by increasing the water permeability up to three times than normal could reduce the energy consumption pressure vessels for seawater desalination about 15% and 44%, respectively.

Technology	Specific energy consumption (kWh/m³)		
	Electric thermal	Thermal	Total electric equivalent
ED	1–3.5	—	1–3.5
EDR	1–2	—	1–2
SWRO	3–6	—	3–6
BWRO	0.5–3	—	0.5–3
MVC	7–15	—	7–15
MD	1.5–4	4–40	3–22
FO	0.2–0.5	20–150	10–68

ED = electrodialysis; EDR = electrodialysis reversal; BWRO = brackish water reverse osmosis; SWRO = seawater reverse osmosis; MVC = mechanical vapor compression; MD = membrane distillation; MSF = multi-stage flash; MED = multiple effect distillation; MEB = multi-effect boiling; FO = forward osmosis.

Table 1.
Specific energy consumption (SEC) by different desalination techniques.

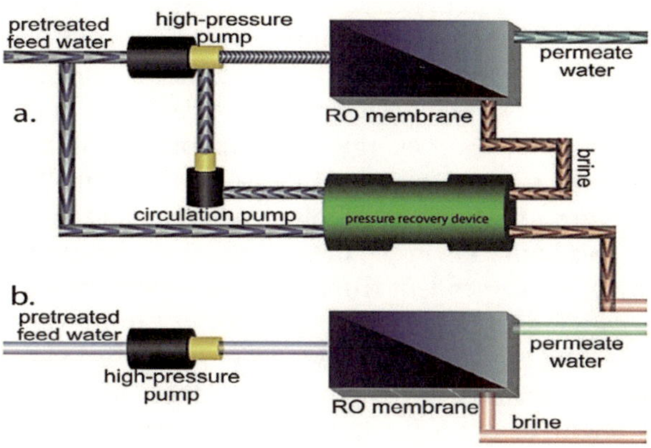

Figure 1.
The RO process diagram with (a) and without (b) pressure recovery for SWRO and BWRO respectively.

In the context of wastewater reclamation, even greater savings (e.g., 45% less energy input and 63% fewer pressure vessels [10]) can be achieved. Moreover, increasing the properties of the membrane selectivity can cause improvement the quality of the product [11].

Recent studies introduce the promise of developing new membrane materials. These materials can desalinate water while showing far greater permeability than traditional reverse osmosis (RO) membranes. But the question remains whether higher permeability means significant reductions in the cost of desalinated water. Research evaluates the potential of ultra-permeable membranes (UPM) to improve the performance and cost of RO.

2.2.1 Ultra-permeable membranes (UPM)

By modeling the mass transport inside a reverse osmosis pressure vessel (PV), the study assesses how much tripling water permeability lowers energy consumption. And also lowers the number of required pressure vessels for a particular desalination plant. The findings were very interesting, it proved that a tripling (3×)

in permeability permits 44% fewer pressure vessels and 15% less energy for a seawater reverse osmosis plant (SWRO) [10, 12]. This is done Â at a both given capacity and recovery ratio. Moreover, tripling permeability results in 63% fewer pressure vessels or 46% less energy for brackish water reverse osmosis (BWRO). However, it also shows that the energy savings of ultra-permeable membranes (UPM) exhibits a law of diminishing returns due to thermodynamics and concentration polarization at the membrane surface [10].

In terms of reducing energy consumption, the benefits of ultra-permeable membranes (UPM) are limited to approximately 15% in the case of SWRO. It also shows that membranes with 3× higher permeability reduces number of pressure vessels by 44% for seawater reverse osmosis RO plants SWRO. And 63% in brackish water RO plants BWRO. This does not affect the energy consumption or permeate recovery [13].

In order to calculation of systems-level quantities the typical RO process diagram that shown in **Figure 2**, is used. In SWRO systems, for pressurizing the feed using mechanical energy Regenerated force from isobaric brine, pressure recovery devices (PRDs) are used (**Figure 2a**), while at BWRO typically this is not done (**Figure 2b**).

In case of energy consumption, ultra-permeable membranes proved to lower energy consumption of seawater reverse osmosis systems—SWRO—by %15. While on the other hand lowered energy consumption of brackish water reverse osmosis systems—BWRO—by 46%. The research was made at the same permeate flow per pressure vessel as what is typical nowadays. As can be shown in **Figure 2a** by reducing the inlet pressure, lower energy consumption (membrane area, feed flowrate and for a given recovery ratio) would be obtained. In SWRO (the line with purple dye in the figure), the pressure of inlet feed reduces to the outlet of the brine osmotic pressure. This limitation in the membrane, that corresponds to the osmotic pressure of the brine, is independent from membrane performance. As can be seen in the **Figure 2a**, with increasing A_m up to triple from 1 to 3 L (m^2 h bar), we can reduce the inlet pressure about 1% and reach from 70 bar to 63 bar. For every 1% reduction in the inlet pressure, the SEC could be reduced up to 1.5%. However, as can be seen in this figure, any further improvements in membrane permeability beyond 3 L (m^2 h bar)$^{-1}$, since 63 bar is already within 1% of the osmotic limit for SWRO at the chosen recovery ratio, would have essentially no effect on energy consumption.

As can be shown in **Figure 2a**, in order to achieve 65% recovery in BWRO and with increasing A_m, inlet pressure rapidly drops. Due to the limitation of the osmotic

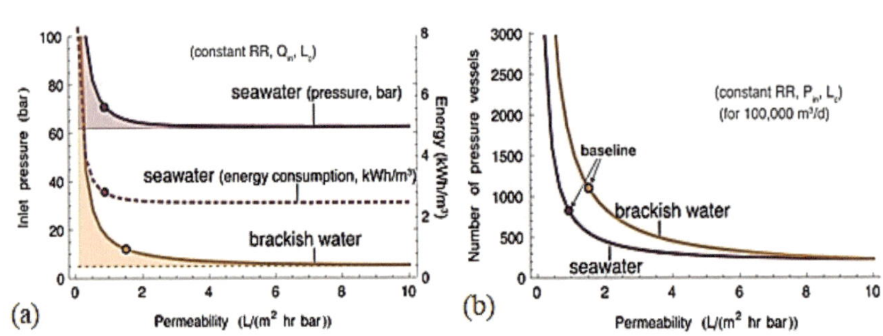

Figure 2.
Investigation of key performance criteria and their effect in membrane permeability for BWRO at 2000 ppm NaCl (orange) and SWRO at 42,000 ppm NaCl (purple). (a) Energy consumption (dashed) and minimum required inlet pressure (solid lines) at fixed feed flowrate and recovery. In BWRO, energy consumption and pressure are linearly related. (b) Number of pressure vessels required for a total capacity of 100,000 m^3 day^{-1} at fixed recovery ratio and pressure. Membrane width is held fixed in both subplots.

for BWRO that is only a fraction of that of SWRO, with the increase in membrane permeability up to triple, it is causing a much greater reduction in inlet pressure, namely down to 6.4 bar from 12 bar in the case of thin-film composite (TFC) membranes (a 46% reduction in pressure and energy consumption). In the membranes with more permeability, with increasing the membrane's water permeability (A_m) (L $(m^2 \text{ h bar})^{-1}$) to over 5 L $(m^2 \text{ h bar})^{-1}$, the pressure essentially reaches the asymptotic limit. So, for the RO plant in the stage of brackish water, the UPMs could reduce the energy consumption to half. In the BWRO, the number of pressure vessels is lower than the SWRO. On the basis of **Figure 2b**, with a tripling A_m, we can reduce the pressure vessels up to 63%, for a given plant capacity, by increasing the feed flowrate per vessel from 139 m^3 day^{-1} to 378 m^3 day^{-1}. Furthermore, increases in feed flowrate have no effect on the energy, since, viscous losses in a BWRO system represent a negligible component of the overall energy consumption [3].

Commercial RO membranes are dominated by TFC polyamide and its derivatives **Figure 3**. These membranes are facing critical challenges such as low selectivity, relatively low water permeability and high fouling tendency [2]. For example, in RO membranes, TFC has a typical water permeability range from ~1–2 L m^{-2} h^{-1} bar^{-1} for SWRO membranes and ~ 2–8 L m^{-2} h^{-1} bar^{-1} for BWRO [10, 14]. So, in synthesizing novel RO membranes, focused on the improvement of separation properties and better antifouling performance that is a key research focus in the field of desalination.

When it comes to capital costs, on the basis of our analysis, we can propose certain qualitative trends. According to Global Water Intelligence, in a typical SWRO plant with capacity of 150,000 m^3 day^{-1}, the levelized capital cost today is about 0.20 $ per m^3 (excluding land) that 20% of this cost is due to piping,, pressure vessels and membranes [15, 16]. So, with using of UPMs membrane, in a surface area similar to conventional membranes but with triple permeability, membranes can be reduced by up to 44%, in this situation the membranes would save on the order of 0.02 $ per m^3 in capital costs. The benefits are more significant for BWRO. in BWRO systems with UPMs membrane, saw that reduction of the energy consumption could be up to 46% [8]. Following increase of membrane permeability mass transfer coefficients and also typical cross-flow velocities decrease. With enhancement of membrane permeability, permeate water flux increases routinely [10].

The consequences of producing a product with less working pressure or more permeability can be estimated with confidence. As described above, the energy savings in SWRO with UPMs membrane could be limited to about 15%. At SWRO plants, because of the high salinity of seawater, operation has been optimized in such a way that these plants work with minimum pressure (60–70 bar) in order to extract permeate water from seawater [8, 10]. The difference between pre- and

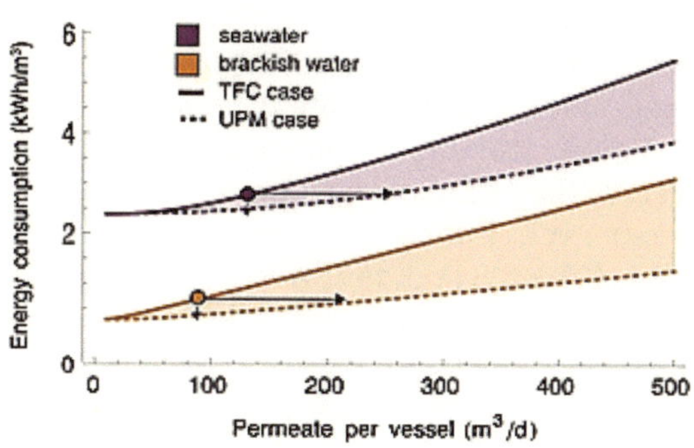

Figure 3.
Ultra-permeable membranes UPM thin film composite TFC for BWRO and SWRO.

post-treatment is about ~1 kWh m^{-3}, in RO stage, a 15% reduction in the energy consumption could only reduce ~10% of the overall cost of the energy in SWRO plants. With the reducing of the total energy consumption in SWRO plants from 3.8 kWh to 3.5 kWh, If the price of electricity is assumed to be 0.10 $ per kWh, could be saved the cost about 0.03 $ per m^3 [17, 18].

Wilf [19] evaluated with replacing the RO elements with membranes which have 80% higher permeability, in situation which recovery ratio and feed salinity was 85% and 1500 ppm, respectively, the SEC of BWRO decrease. He found that in two different averages flux (25.5 LMH and 34 LMH) the SEC was decreased (from 0.52 to 0.40 kWh/m^3 and from 0.72 to 0.49 kWh/m^3, respectively).

Franks et al. [20] evaluated, in BWRO plants, when a membrane element with 34.1 m^3/d of permeate flow replace with another elements that has 45.4 m^3/d of permeate flow, the SEC decrease. In this study, with decreasing the feed pump pressure 9.8–8.3 bar, the specific energy consumption decreased from 0.41 to 0.35 kWh/m^3 (the pump efficiency was 83%, the recovery ratio was 85% and the feed salinity was 1167 ppm (for wastewater). The simulation conditions were shown in **Tables 2** and **3**.

For a BWRO plant, Werber et al. [24] assumed a 85% recovery rate and feed with NaCl concentration about 5844 ppm. They observed, in a single-stage process, with increasing the water permeability in membrane from 4 to 10 LMH/bar, the SEC can be reduced up to 2.2%. On the other hand, in this study observed that in a two-stage RO with membrane permeability of about 4 LMH/bar, the required energy was 22% lower (0.11 kWh/m^3) than the single-stage RO, also the SEC decreased by increasing the membrane permeability from 4 to 10 LMH/bar by 12% (0.05 kWh/m^3) that compare to a single-stage BWRO was slightly larger. In this study, in SWRO with single stage process and membrane permeability about 2 LMH/bar, the hydraulic pressure was only 7.6% above the brine osmotic pressure (**Figures 4** and **5**). The results of their findings of the relationship between membrane water permeability and the SEC have shown.

Busch et al. [29] assessed the CAPEX and OPEX reductions with higher permeable SWRO elements. They compared the energy use, power cost, water cost by replacing SW30HR-380 with 28.4 m^3/d of permeate flow rate and 99.75% of NaCl rejection rate by SW30HR LE-400 with 34.1 m^3/d of permeate flow rate and 99.70% of NaCl rejection rate using the test results for each element. Test conditions and calculation assumptions were 32,000 mg/L NaCl of feed concentration, 8% of recovery rate, 55 bar of feed pump pressure, 5 years of operating time, 20% of RO membrane elements replacement rate per year, 90% of pump efficiency, and 0.08 US$/kWh of power cost. The pretreatment, chemical cleaning, and other costs were not considered. They indicated that decreasing membrane area by using higher water permeability RO elements can decrease the water cost by 4.7% from 0.190 to 0.181 US $/m^3 with the same energy cost.

For SWRO, the energy cost contributes 40–50% of the total water production cost; therefore, the ratio of the specific membrane cost to the total water production cost is about 1.2–6%. Hence, doubling the membrane water permeability halves the specific membrane cost so that the total water production cost is reduced to 0.6–3%. When the cost of pressure vessels is taken into consideration, the decrease of total water production cost is 0.7–3.5% [12]. But, with increasing the membrane permeability, the feed velocity and the pressure loss increase, as a result, more energy is needed, these could increase the SEC up to 6%.

As can be shown in **Figure 6**, Cohen-Tanugi et al. [10] calculated the total number of pressure vessels needed in a single-stage SWRO and BWRO with 100,000 m^3/d permeate and 42,000 ppm and 2000 ppm salinity concentration, respectively.

Condition							Author (year)	Reference
Feed concentration	Recovery rate	Average system flux/average TMP	No. of elements per vessel	Salt rejection	Pump efficiency	ERD efficiency		
ppm or mg/L	%	LMH/bar	—	%	%	%		
35,000 mg/L TDS	50	–/15.5	N.D.	99	100	100	Zhu et al. (2009)	[21]
42,000 ppm NaCl	42	16/–	8	99.8	75	97	Cohen-Tanugi et al. (2014)	[10]
32,000 ppm NaCl	50	N.D	N.D.	N.D.	85	95	Shrivastava et al. (2015)	[22]
30,000 mg/L NaCl (π_F = 25.6 bar)	50	15/–	Not considered	100	100	100	McGovern et al. (2016)	[23]
35,000 mg/L NaCl	50	15/–	N.D	*B*-value was used	100	100	Werber et al. (2016)	[24]
35,000 mg/L NaCl	50	22.9/–	8	100	100	100	Mazlan et al. (2016)	[25]
35,000 mg/L NaCl	N.D.	N.D.	N.D	100	N.D	N.D	Shi et al. (2017)	[26]
35,000 ppm NaCl	70	15/–	8	100	100	100	Wei et al. (2017)	[27]
40,000 ppm	50	N.D.	N.D	N.D	85	95	Karabelas et al. (2018)	[28]

This article was published in [12], Page 6, Copyright © 2021 Elsevier B.V (ScienceDirect) (2019).

Table 2.
Simulation conditions of each reference that includes the relationship between membrane water permeability and SEC for SWRO [12].

Condition							Author (year)	Reference
Feed concentration	Recovery rate	Average system flux/average TMP	No. of elements per vessel	Salt rejection	Pump efficiency	ERD efficiency		
ppm or mg/L	%	LMH/bar	—	%	%	%		
3500 mg/L TDS	50	−/1.55	N.D	99	100	100	Zhu et al. (2009)	[21]
2000 ppm NaCl	65	13.2/−	7	99.8	75	97	Cohen-Tanugi et al. (2014)	[10]
804 mg/L TDS	85	N.D.	N.D	N.D	85	95	Shrivastava et al. (2015)	[22]
5844 mg/L NaCl	75	15/−	N.D	*B*-value was used	100	100	Werber et al. (2016)	[24]
1000 mg/L NaCl	N.D.	N.D.	N.D	100	N.D	N.D	Shi et al. (2017)	[26]
3000 ppm NaCl	60–98	15	8	100	100	100	Wei et al. (2017)	[27]
2000 ppm	70	N.D	N.D	N.D	85	95	Karabelas et al. (2018)	[28]

This article was published in [12], Page 7, Copyright © 2021 Elsevier B.V (ScienceDirect) (2019).

Table 3.
Simulation conditions of each reference that includes the relationship between membrane water permeability and SEC for BWRO [12].

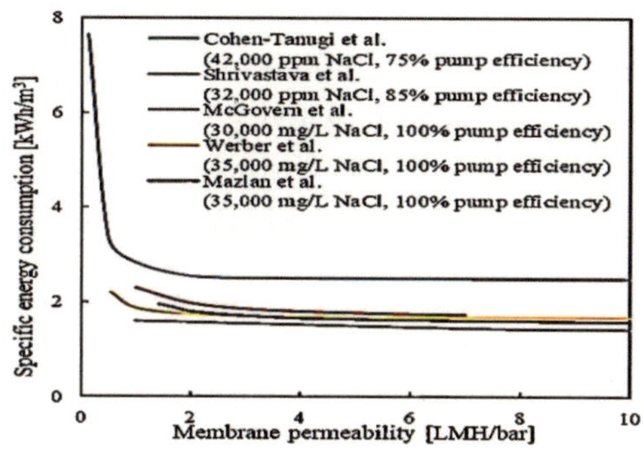

Figure 4.
Calculated specific energy consumption as a function of membrane water permeability for single-stage SWRO from several references at different conditions.

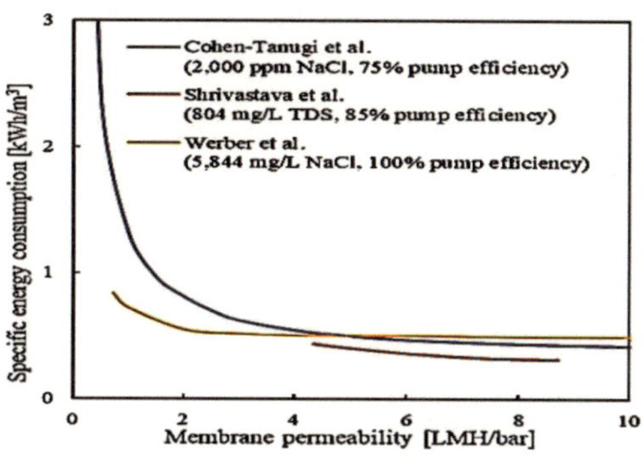

Figure 5.
Calculated SEC as a function of membrane permeability for BWRO from several references.

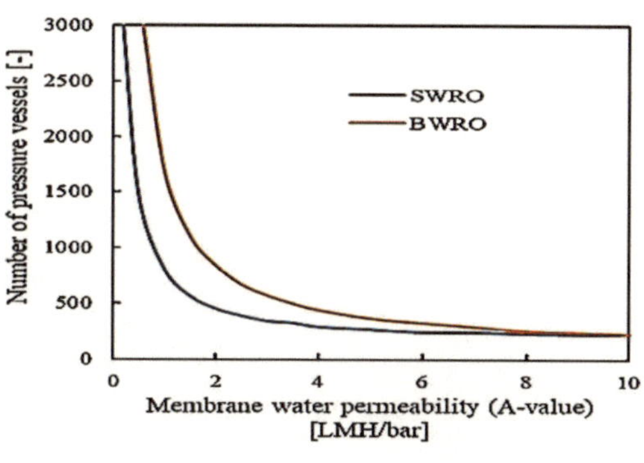

Figure 6.
Calculation of the required number of pressure vessels in order to investigation of function of the membrane water permeability for single-stage BWRO with12 bar feed pump pressure, 2000 ppm feed, 65% recovery rate and single stage SWRO with 42% recovery rate, 42,000 ppm feed and 70 bar feed pump pressure for total capacity of 100,000 m^3/d.

3. RO membrane: types, structures and materials

Based on the membrane structure, The RO membrane is consisted of two groups: conventional thin-film composite and thin-film nanocomposite. Based on the thin-film material, conventional RO membrane is classified into two main groups: cellulose acetate (CA) and aromatic polyamide (PA). The RO membrane on the basis of the membrane configuration can be divided into three main groups: hollow-fiber, flat-sheet (plate-and-frame) and spiral-wound [30, 31].

3.1 Conventional thin-film composite membrane structure

The RO membrane which is used widely today are composed a semipermeable thin film (0.2 um), made of either CA or PA, supported by a 0.025- to 0.050-mm microporous layer that in turn is cast on a layer of reinforcing fabric (**Figure 7**). Maintaining and reinforce the membrane structural integrity and durability is the main functions of the two support layers underneath the thin film [31].

In the dense semipermeable polymer film that is made up from a random molecular structure (matrix), there is no any pores. Water molecules are transported through the membrane film by diffusion and travel on a multidimensional curvilinear path within the randomly structured molecular polymer film matrix [12, 31].

3.2 Thin-film nanocomposite membrane structure

Thin-film nanocomposite (TFC) consisting from two main structure; inorganic nanoparticles in traditional membrane polymeric film structure (**Figure 8**) and highly structured porous film consisting of a densely packed array of nanotubes (**Figure 9**). In **Figure 8**, part A shows the thin film of a conventional PA membrane that supported by the polysulfone support layer. Part B shows the same type of membrane with embedded nanoparticles.

In nanocomposite membrane the specific water permeability, at comparable salt rejection, is higher than the conventional RO membrane. In addition, the fouling rates in TFC membrane, at the same operation conditions, is lower in comparison to conventional TFC RO membrane. In other words, in case of production of tubular membranes with completely uniform size, theoretically the membrane could produce up to 20 times more water per unit surface area than the common RO membrane commercially available on the market today.

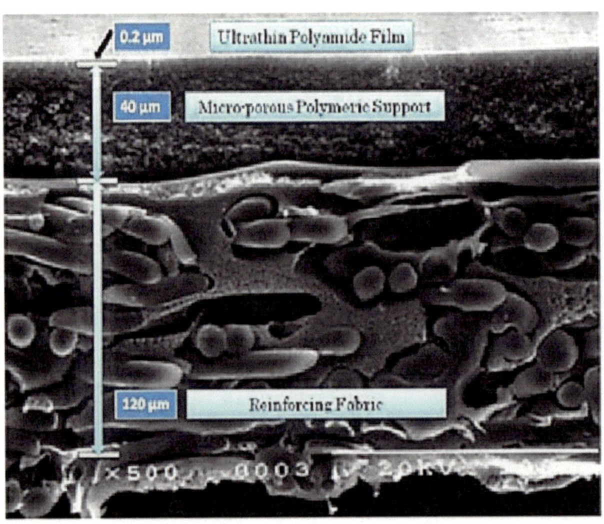

Figure 7.
Structure of a typical reverse osmosis RO membrane with ultrathin PA film.

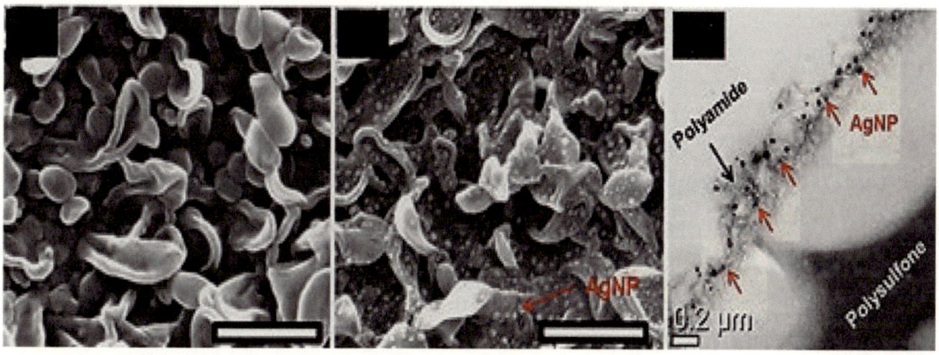

Figure 8.
Polyamide reverse osmosis RO membrane with nanoparticles.

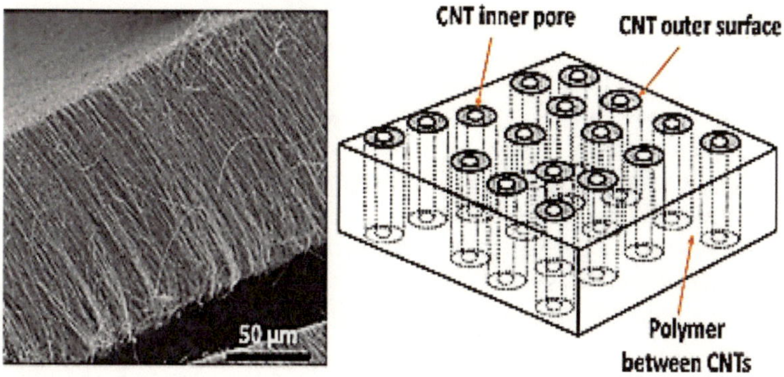

Figure 9.
The RO membrane with carbon nanotubes [32].

3.3 Cellulose acetate CA membrane

For the first time in the late 1950s the thin semipermeable film as the first membrane element from cellulose acetate (CA) polymer was made at the University of California, Los Angeles [33]. Although the CA membrane is similar to the aromatic polyamide (PA), but, because of the existence of the top two layers (the ultrathin film and the microporous polymeric support) in the main structure of the CA that are made of different forms of the same CA polymer, the CA is different from PA [34]. In PA membrane unlike the CA these two layers consist of two completely different polymers, the polyamide and polysulfone form the semipermeable films and microporous supports, respectively. In CA membrane similar to PA membrane, thickness of the film layer is typically about 0.2 μm, but the thickness of the entire membrane in CA membrane is different (about 100 μm) from the PA membrane (about 160 μm) [35].

One of the important advantages of CA membrane is its surface very little charge, which is considered practically uncharged, while in PA membrane, because of negative charge in the surface of the membrane, with use of cationic polymers for water pretreatment, the potential for fouling increases dramatically. Furthermore, due to the smoother surface in CA membrane than the PA membrane, the CA membrane less clogged [34].

Some disadvantages of the CA membrane are; low operation temperatures 35°C (95°F) and narrow pH working range (4–6). Operation outside of this pH range can cause hydrolysis of the membrane, also, exposure to temperatures above 40°C (104°F) causes membrane compaction and failure [33]. Due to these limitations, the pH in feed water interring to the CA membrane has to be reduced and maintain between 5 and 5.5, which, in order to normal plant operation, the use of acid increases. in addition, the requires reverse osmosis RO permeate adjustment by addition of a base (typically sodium hydroxide) to achieve adequate boron rejection [36].

Since CA membrane has a higher density than PA membrane, it creates a higher head loss and has to be operated at higher feed pressures, which results in increase in energy consumption. Despite their disadvantages, due to their high tolerance to oxidants (chlorine, peroxide, etc.) than the PA membrane, CA membrane is used in municipal applications for ultrapure water production in pharmaceutical and semiconductor industries and for saline waters with very high fouling potential (mainly in the Middle East and Japan).

3.4 Aromatic polyamide membrane

The aromatic polyamide (PA) membrane widely used in RO membrane structure and production of potable and industrial water at today. The thin polyamide film of this type of semipermeable membrane is formed on the surface of the microporous polysulfone support layer. For production of PA membrane uses the interfacial polymerization of monomers containing polyamine and then immersion of it in the solvent containing a reactant to form a highly cross-linked thin film. Because of some properties such as lower working pressure, lower salt passage than CA membrane and higher productivity (specific flux), the PA membranes have wider application at today [37, 38].

By changing pH, the surface charge of PA and CA membrane is also changes. For example, CA membrane has a neutral charge while, PA membrane in pH greater than 5 has a negative charge, and for this reason, co-ion repulsion amplified and therefore salt rejection is higher than CA membrane. However, when pH is lower

Parameter	Polyamide membrane PA	Cellulose acetate CA membrane
Salt rejection	High (>99.5%)	Lower (up to 95%)
Feed pressure	Lower (by 30–50%)	High
Surface charge	Negative (limits use of cationic pretreatment coagulants)	Neutral (no limitations on pretreatment coagulants)
Chlorine tolerance	Poor (up to 1000 mg/L-hours); feed dechlorination needed	Good; continuous feed of 1–2 mg/L of chlorine is acceptable
Maximum temperature of source water	High (40–45°C; 104–113°F)	Relatively low (30–35°C; 86–95°F)
Cleaning frequency	High (weeks to months)	Lower (months to years)
Pretreatment requirements	High (SDI < 4)	Lower (SDI < 5)
Salt, silica, and organics removal	High	Relatively low
Biogrowth on membrane surface	May cause performance problems	Limited; not a cause of performance problems
pH tolerance	High (2–12)	Limited (4–6)

Table 4.
Comparison between polyamide PA membrane and cellulose acetate CA membrane [37].

than 4, the charge of the PA membrane changes to positive and rejection reduces significantly to lower than the CA membrane [38]. One another of the most important advantage of the PA membrane is much wider operation pH range (2–12). This allows easier maintenance and cleaning. Furthermore, the PA membrane has resistant to biodegradation and have a longer useful life (5–7 years) compare to usually membrane (3–5 years). From Aromatic polyamide membrane is used in order to production of membrane elements for nanofiltration, seawater desalination and brackish water [33, 37].

3.5 Comparison between PA and CA membrane

For PA membrane, the chlorine is and other strong oxidants the biggest threat and can destroying the membrane structure and consequently reduce the salt rejection performance of the membrane. In order to biofouling control in nanofiltration and RO membranes, Oxidants are widely used, so, before separation, the feed water to PA membrane has to be dechlorinated. In **Table 4**, the key parameters of polyamide and cellulose acetate RO membrane has been shown.

4. Recent development of novel membranes for desalination

In commercial RO membranes, almost the majority of materials that are used are dominated by thin-film composite (TFC) polyamide and its derivatives. At these membranes, we are faced with critical challenges like relatively low water permeability, high fouling tendency and low selectivity [39]. For example, in commercial TFC RO membranes the typical water permeability for seawater reverse osmosis (SWRO) and brackish water reverse osmosis (BWRO) is range from $\sim 1-2\ L\ m^{-2}\ h^{-1}\ bar^{-1}$ and $\sim 2–8\ L\ m^{-2}\ h^{-1}\ bar^{-1}$, respectively [40]. One of the fields in desalination that is been focus on it, is synthesizing novel membranes with better antifouling performance and improved separation properties.

Much of the exciting progresses are fueled by the recent emergence of promising novel materials for desalination. Among them, the most notable examples include aquaporin (AQP) proteins [11, 41, 42] and some carbon-based materials such as carbon nanotubes (CNTs) [43] and graphene-based materials [44]. At the moment, in RO membranes, the old asymmetric cellulose acetate largely replaced with TFC polyamide membranes [45, 46]. New TFC polyamide membranes compared to the former membranes, have been shown better performance in water permeability and salt rejection (e.g., in SWRO rejection of NaCl is >99.9%), pH tolerance (1–11) and wider operating temperature range (0–45°C) [11].

4.1 Novel materials and methods for synthesizing desalination membranes

4.1.1 Carbon-based materials

Because of exceptional water transport properties of Carbon based materials (CBMs), e.g., nanoporous graphene (NPG) [47, 48], carbon nanotubes (CNTs) [49, 50], and graphene oxide (GO) [11, 51] have been raised hopes of improvement in the membrane processes (**Tables 5** and **6**). In these materials, the characteristic of water channel dimensions as well as chemical modifications (e.g., the presence of carboxyl, amine and other groups) determines the rejection properties [11, 56, 75]. The characteristic of the channel dimensions in NPG and CNTs are sorted by their respective pore sizes [51]. In CNTs and NPG, the channel sizes determined by their synthesis conditions, but, in GO the characteristic channel size is highly dependent

	Polyamide	AqpZ	CNT	NPG	Graphene oxide (GO)
Material transport mechanism	Cross-linked polymer solution-diffusion	Natural protein for charge repulsion and size-exclusion	Material with 1-D carbon size-exclusion (enhanced by charge repulsion)	Material with 2-D carbon size-exclusion (enhanced by charge repulsion)	material with 2-D carbon size-exclusion (enhanced by charge repulsion)
Characteristic channel size (Å)	Irregular pores in a random network, characteristic pore diameter of ~4–5.8a Å based on positron annihilation lifetime spectroscopy [52], possibly heterogeneous pore distribution for some membranes [53]	Well-defined hour-glass-shaped channel [54], pore size of ~3 Å [55]	Well-defined cylindrical pores (e.g., ~13–20 Å [56])	Nano-sized pores across 1-atomthick graphene layer, possibly with non-uniform pore sizes (e.g., obtained from plasma etching, ~5–10 Å [52])	Channels formed by adjacent GO layers, channel size depending on the degree of oxidation or solution environment [57]
Separation properties	~1–2 $L\ m^{-2}\ h^{-1}$ bar-1 for SWRO and ~ 2–8 $L\ m^{-2}\ h^{-1}\ bar^{-1}$ for BWRO; ~ > 99% NaCl rejection (obtained from cross-flow filtration tests) [40]	~600 $L\ m^{-2}\ h^{-1}\ bar^{-1}$; nearly 100% NaCl rejection (obtained from stopped-flow measurements of AQP-containing vesicles) [41, 58]	Gas permeability is >10 times higher than the predictions of the Knudsen diffusion model; experimental water permeability is >1000 times higher than the calculated results from continuum hydrodynamics (obtained from measuring the water flux of an aqueous suspension of gold nanoparticles; CNTs pore density ≤ $2.5 \times 10^{11}\ cm^{-2}$ and length of ~3 μm) [56]	~3.6 × 106 $L\ m^{-2}\ h^{-1}$; nearly 100% KCl rejection at 40°C for a 5-μm-diameter sample (obtained from gravity-driven test in an oven) [47]	Water permeability is at least 10^{10} times faster than that of helium (obtained from weight-loss measurements by a 1-μm-thick GO membrane) [44]; water permeability and rejection are sensitive to the interlayer spacing
Antifouling properties	Prone to fouling [36]	Not reported in literature	Antimicrobial [56] (and improved hydrophilicity for functionalized CNTs [59])	Not reported in literature	Antiadhesion (due to hydrophilicity) and antimicrobial [60]
Electrical conductance	No	No	Yes	Yes	No

This article was published in [11], Page 5, Copyright © 2021 Elsevier B.V (ScienceDirect) (2018).

Table 5.
Material properties of polyamide, AqpZ, CNT, NPG and graphene oxide [11].

Type	Classification	P_w ($L\ m^{-2}\ h^{-1}\ bar^{-1}$)	Rejection (%)	Testing conditions and membrane area (cm^2)	Results	Ref.
PRL AqpZ DOPC[a]	NF	3.6	R_{NaCl} = 20%	1 mM NaCl @1 bar Area: 28.3	DOTAP coated NF270, with both decreased water flux and R_{NaCl} compared to virgin membranes	[61]
AqpZ-ABA[b]	NF	34.2	R_{NaCl} = 32.9%	200 ppm NaCl @5 bar Area: 0.071	Silanized CA substrate, high P_w with low R_{NaCl}, the amount of AqpZ has huge impact on membrane performance	[62]
AqpZ-ABA	NF	16.1	R_{NaCl} = 45.1%	200 ppm NaCl @5 bar Area: 0.2	Gold coated porous alumina substrate cross-linked with disulfide: high P_w with less defects	[63]
AqpZ-DOPC/ DOTAP[c]	NF	5.5	R_{NaCl} = 75% R_{MgCl2} = 97%	500 ppm NaCl @4 bar Area: 19.56	AQP containing lipid bilayers deposited on PSS/PEI/ PAN substrate	[64]
AqpZ-ABA	FO	J_v[d] = 16.4 $L\ m^{-2}\ h^{-1}$	R_{NaCl} = 98.8%	0.3 M sucrose as DS, 200 ppm NaCl as FS[e] Area: 0.096	Gold and cysteamine coated polycarbonate with UV cross-linking	[65]
AqpZ-DOPC/ DOTAP	FO/NF	J_v = 23.1 $L\ m^{-2}\ h^{-1}$ NF:6.31	FO: Js = 3.1 g $m^{-2}\ h^{-1}$ NF: R_{MgCl2} = 90%	2 M $MgCl_2$ as DS, DI water as FS 2000 ppm $MgCl_2$ @ 4 bar Area: 36	AqpZ-DOPC/DOTAP coated on PDA modified porous polysulfone substrate via amidation reaction to form covalent bonds.	[66]
TFN AqpZ-DOPC	RO	4	R_{NaCl} = 97% @ 5 bar	10 mM NaCl @5 bar Area: > 200	AqpZ containing vesicles incorporated in PA layer serving as protection layer via IP. Large membrane area can be obtained	[67]
AqpZ-DOPC	RO	8	R_{NaCl} = 97.5%	500 ppm NaCl @5 bar Area: 34.2	Vesicles embedded in PA rejection layer with superior water flux	[68]
AqpZ-DOPC	RO	4.1	R_{NaCl} = 97.2%	10 Mm NaCl @10 bar Area: 42	Vesicles embedded in PA rejection layer for long term stability test	[69]
AqpZ-POPC/ POPG/cholesterol[g]	NF	~6	R_{MgCl2} = 96%	200 ppm $MgCl_2$ @ 4 bar Area: 0.785	Vesicles embedded in PSS/PAA LBL[f]. Membranes with AqpZ showed Pw ↑ 60% with $MgCl_2$ rejection↑ compared to the control	[70]
AqpZ-DOPC	NF	36.6	R_{MgCl2} = 95%	100 ppm $MgCl_2$ @ 1 bar Area: 28.3	PDA coated vesicles incorporated in cross-linked PEI matrix	[71]

Type	Classification	P_w ($L\ m^{-2}\ h^{-1}\ bar^{-1}$)	Rejection (%)	Testing conditions and membrane area (cm^2)	Results	Ref.
AqpZ-ABA	NF/FO	NF: 22.9 $J_v = 5.6\ L\ m^{-2}\ h^{-1}$	$R_{NaCl} = 61\%$ $R_{MgCl2} = 75\%$ FO: $R_{NaCl} = 50.7\%$	200 ppm salt @5 bar 0.3 M sucrose as DS and 200 ppm NaCl as FS	AqpZ-vesicle loaded membrane cross-linked by UV	[72]
AqpZ-ABA	FO	$J_v = 43.5\ L\ m^{-2}\ h^{-1}$	$J_s = 8.9\ g\ m^{-2}\ h^{-1}$	0.5 M NaCl as DS, DI water as FS Area: 0.196	Pressure assisted sorption, further coated with cysteamine and cross-linked by polydopaminehistidine. The control membrane has FO water flux of 8.6 $L\ m^{-2}\ h^{-1}$ and $Js = 6.6\ g\ m^{-2}\ h^{-1}$	[73]
AqpZ-POPC/ POPG/Cholesterol	FO	$J_v = 21.8\ L\ m^{-2}\ h^{-1}$	$Js = 2.4\ g\ m^{-2}\ h^{-1}$	0.3 M sucrose as DS and 200 ppm $MgCl_2$ as FS Area: 0.785	Magnetic-assisted AQPs embedded membranes	[74]

This article was published in [11], Page 8, Copyright © 2021 Elsevier B.V (ScienceDirect) (2018).
[a]*DOPC: 1,2-dioleoyl-sn-glycero-3-phosphocholine.*
[b]*ABA: methacrylate end functionalized poly(2-methyloxazolineb-dimethylsiloxane-b-2-methyloxazoline) PMOXA(1000)-b-PDMS(4000)-PMOXA(1000) triblock.*
[c]*DOTAP: 1,2-dioleoyl-3-trimethylammonium-propane.*
[d]*J_v: FO water flux; J_s: FO solute flux.*
[e]*DS: draw solution; FS: feed solution.*
[f]*LBL: layer by layer deposition of polyacrylic acid (PAA) and polystyrene sulfonate (PSS).*
[g]*POPC: 1-palmitoyl-2-oleoyl-sn-glycero-3-phosphocholine; POPG: 1-palmitoyl-2-oleoyl-sn-glycero-3-phospho-(19-rac-glycerol).*

Table 6.
Summary of RO, NF and FO performance of biomimetic membranes [11].

on solution environment and its degree of oxidation [11]. In this section, we have summarized the detailed materials properties of NPG, CNT and GO [76–78].

5. Hybrid technologies: the future of energy efficient desalination

Desalination processes traditionally rely on mechanically driven membrane processes such as reverse osmosis (RO) or thermal distillation such as multi-effect distillation (MED) and multi-stage flash (MSF). In the use of membrane technologies, the principle is based on the use of technology with easy operation, limited use of chemicals, compactness, low energy consumption and the development of enhanced membrane materials [79]. Some emerging desalination technologies like forward osmosis (FO) and freeze desalination (FD), despite the serious challenges in the road to commercialization, have also recently garnered interest.

In a desalination plant, roughly 20–30% of the overall cost in water production is related to the energy [22, 29]. There is growing interest in combining the benefits of two or more systems, to meet specific water quality goals and/or reduce energy consumption. Using hybridization in desalination technologies is often in order to one or more objectives such as increasing water recovery rate, eliminating the need for a second pass or reducing brine salinity. Hybrid systems have been considered as economically superior alternatives to standalone systems due to their ability to reduce energy consumption and therefore cost of desalinated water through improved recovery rate and/or water quality [80].

5.1 Current status and energy consumption in desalination systems

5.1.1 Multi-stage flash (MSF)

The basis of working multi-stage flash distillation (MSF) is distills sea water by flashing a portion of the water into steam in multiple stages of what are essentially countercurrent heat exchangers [81]. In order to occur the flashing, the pressure in each stage must be lower than the vapor pressure of the heated liquid. by passing the cold feed from each stage, it be heated that is further heated in the brine heater. At the time of brine flows return, because of higher temperature than the boiling point in brine, in the normal pressure, a fraction of the brine boils to the steam. After this stage, the steam is starting to condensation on the external surface of heat exchanger tubes [82]. At the moment, two more well-known configurations of the MSF are the once-through MSF (MSF-OT) and brine mixing MSF (BM-MSF) [80].

At this moment, about 23% of all desalinated water in the world is produced by MSF plants, but due to the high energy consumption, their use is declining [83].

In the practical scale, for commercial MSF systems, a value of 8 to 12 $kg_{distillate}/kg_{steam}$ are typically reported [84]. Some parameters like as corrosion and pipe fouling, scale formation and etc. reduce the energy efficiency of MSF systems. In MSF plants the amount of energy that consume is between 23 and 27 kWh/m^3 [80, 85]. El-Naser [86] reported that in MSF plants the energy consumption is average 12–24 kWh/m^3.

5.1.2 Multi-effect distillation (MED)

One of the oldest industrial desalination processes that are used today is Multi-effect distillation (MED) [87, 88]. The MED evaporator consists of cells, called effects, decreasing pressure and temperature from first to last, with temperature typically between 65 and 90°C [89]. Each effect consists of evaporator tube bundles on which seawater is sprayed. Heating steam or hot water through the tubes is supplied in the first effect and it transfers energy to the seawater in each effect, causing partial evaporation [90]. In each effect, the low pressure and temperatures affect the boiling point of water and with decreases of its, water becomes evaporate [88]. By using a heat exchanger and condensing the steam, clean distillate water is produced. This product water is pumped into a storage tank while the brine is pumped back into the sea.

For the production of water in a MED plant with a capacity range between 5000 and 50,000 m^3/day, we require thermal energy between 145 and 230 MJ/m^3, which will be equal to 12.2–19.1 kWh/m^3 of electrical energy. Furthermore, for pumps consumption will have been needed 2–2.5 kWh/m^3 of additional electrical energy [91]. Vapor flow and feed configurations are two major parameters that can effect on energy consumption in the MED process.

5.1.3 Electrodialysis (ED)

Electrodialysis (ED) is an electro membrane process in which with use of an electric field ionic and non-ionic components are removed [29]. In these kinds of processes, Anions and cations migrate towards the positive and negative electrode, respectively, and so the separation process happens. As can be show in the **Figure 10**, an ED system consists of alternately arranged anion exchange membranes (AEM) and cation exchange membranes (CEM).

The energy consumption in ED strongly depends to the salt concentration in feed solution. The rate of salt removal is proportional to the electric current [80, 92]. In order to efficient separation of ions from feed solution with high concentration, would require a high potential difference, thus, the use of ED process for seawater desalination, due to high concentration of ions in seawater and the need for high energy consumption, it is not affordable. This process is suitable for solutions with low-concentration of TDS (<5000 mg/L) such as brackish water [93]. Other parts that consume energy is the pumping unit and electrodes. On the basis of recent study, about 1–3% of the total energy consumption is related to these sections [92, 94].

Theoretically, in ED, for producing water with TDS about 800 mg/L the requirement of energy is 3.3 kWh/m^3 and 26 kWh/m^3 for desalination of brackish water and seawater, respectively [95]. On average, 0.7 kWh for each 1000 mg/L

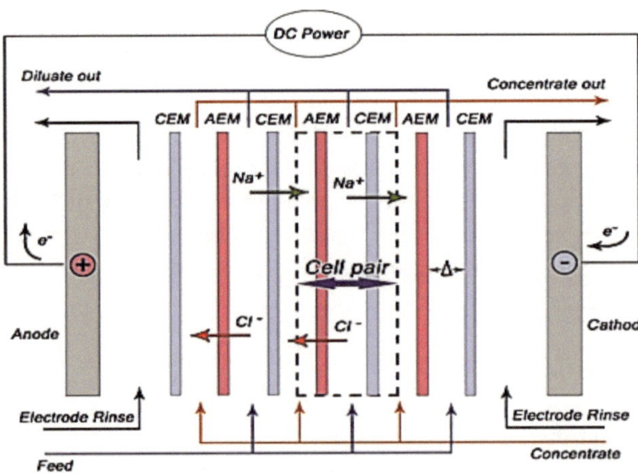

Figure 10.
Schematic of electrodialysis desalination.

TDS removed, 0.5–1.1 kWh/m^3 for pumping, and roughly 5% accounts for energy losses in a brackish water ED desalination system [96]. In a study that was reviewed by Sajtar and Bagley, they found in order to removal of TDS up to 2000 mg/L in feed stream, the energy consumption is ranges from 0.1 to 1 kWh/m^3 [92, 94]. Although ED is typically applied as a room temperature process, introducing a temperature gradient or increasing the temperature of the system can cause energy reductions [94]. Benneker et al. [97] found that the energy required for ED can be reduced by 9% if the temperature of one of the feed streams is increased by 20°C. Increasing the temperature increases ion mobility, reduces electrical resistance of the solution and decreases solution viscosity.

On the basis of the water salinity, the consumption of the electrical energy by an ED system can be about 0.5–10 kWh/m^3 [98]. For example, to lower TDS from 1500 ppm to 500 ppm, an ED unit would consume ~1.5 kWh/m^3. Due to high energy consumption in ED systems, in order to management and reduced the energy consumption, Recently, multi-stage electrodialysis systems have been investigated. Chehayeb et al. [99] found that by using a two-stage system for brackish water desalination the energy consumption can be reduced up to 29%, that, this can reduce the fixed costs. The application of ED remains limited by the high cost of ion exchange membranes and electrodes, and the electrically-driven degradation of polymeric membranes [100].

5.1.4 Membrane distillation (MD)

Membrane distillation is one kind of separation process which in it, a porous membrane with hydrophobic properties is in contact with aqueous heated feed solution on one side. In MD process, the membranes that was use it works like this, that inhibit from the passage of the liquid water, but on the contrary allowing permeability for free water molecules and thus, for water vapor. These membranes are made of hydrophobic synthetic material (e.g. PTFE, PVDF or PP) and offer pores with a standard diameter between 0.1 and 0.5 μm (3.9×10^{-6} and 1.97×10^{-5} in) [80, 101].

Due to the high amount of energy consumption and as a result the high cost of water production, MD has not still achieved widespread commercial implementation in desalination. There are four basic MD configurations included [102, 103];

- direct contact membrane distillation (DCMD).
- vacuum membrane distillation (VMD).
- air-gap membrane distillation (AGMD).
- sweeping gas membrane distillation (SGMD).

In several studies it has been reported that both AGMD and VMD have greater thermal energy efficiency compared to other configurations, which makes them more popular choices for companies seeking to commercialize MD processes. In **Table 7**, the SEC values for several selected MD systems have been reported [102, 116–118].

5.1.5 Forward osmosis

One kind of osmotic process is called forward osmosis (FO) that, in this process, like RO, in order to the separation of water from dissolved solutes, uses a semi-permeable membrane. This process for creating the driving force for separation uses the osmotic pressure gradient, such that a "draw" solution of high

Configuration	Membrane characteristics	Operating conditions		Feed type	SEC (kWh/m^3)	Plant capacity (m^3/h)	Refs.
		T_f (°C)	T_p (°C)				
DCMD	Spiral wound PTFE (SEP GmbH), pore size 0.2 μ, porosity 80%	35–80	5–30	Radioactive solution	6000–1000	0.05	[104]
AGMD	PTFE, pore size 0.2 μ	60–85	—	Seawater	140–200	0.2–20	[105]
AGMD		313–343	—	Brackish water	30.8		[106]
AGMD	PTFE, pore size 0.2 μ, porosity 80%	—	—	Seawater	200–300	3.46–19	[107]
DCMD in hybrid systems	PP models from Microdyn Nadir, Pore size 0.2 μ, porosity 73%	—	—	Seawater	1.6–27.5	931 (overall)	[108]
DCMD	Commercial membranes from membrane with pore size 0.2 μ and thickness 91 μ	39.8–59	13.4–14.4	Distilled water	3550–4580	—	[109]
VMD	PP, thickness 35 μ, pore size 0.1 μ	15–22	—	Underground water	8100.8–9089.5	2.67–6.94	[110]
AGMD	LDPE, thickness 76 μ, pore size 0.3 mμ, porosity 85%, A_m 7.4 m^2	50–70	—	Tap water, synthetic seawater	~65 to ~127	—	[111]
VMD	Flat sheet PP, thickness 400 μm, Pore size 0.1 μ, porosity 70%, A_m 5 m^2	80	—	Distilled water	130	—	[112]
DCMD	PVDF hollow fiber, thickness 240 μm	80	30	Simulated reverse osmosis brine	~130–1700	—	[113]
DCMD	PTFE with PP support, mean pore size 0.5 ± 0.08, porosity 91 ± 0.5, active layer thickness 46 ± 1 μm, A_m 0.67 m^2	60	18–21	Wastewater	1500	3.85	[114]
DCMD	Several commercial membranes with different characteristics	85	20	Seawater	697–10,457	—	[115]

This article was published in [80], Page 9, Copyright © 2021 Elsevier B.V (ScienceDirect) (2020).

Table 7.
Specific energy consumption (SEC) of selected MD systems [80].

concentration is used to induce a net flow of water through the membrane into the draw solution, thus effectively separating the feed water from its solutes [80, 119]. As a result, separation in FO requires little or no hydraulic pressure as a concentrated draw solution (DS) with a greater osmotic pressure draws in water molecules from the feed solution through a membrane [120].

FO is widely promoted as a low-energy desalination technique. For the determination of the energy consumption in these kinds of plants, a DS recovery step is used. During the osmosis step, in order to overcome dropping the pressure in the feed channel, at 50% water recovery, a low-pressure pump is needed, and the energy consumed is equal to ~0.10–0.11 kWh/m^3 [25, 121]. For the osmosis step the values of 0.2–0.55 kWh/m^3 have also been reported [122]. Moon and Lee suggest, in a FO desalination plant, for solute regeneration, the energy consumption range is from 3 to 8 kWh/m^3 [123].

5.2 Hybrid desalination technologies

In a hybrid desalination system in order to reduce costs or enhance performance in compared to individual components, uses from integration of two or more desalination systems. Due to the high cost of investing in hybrid systems, one of the important parts of these kinds of processes is the optimization of hybrid configurations [80].

5.2.1 Electrodialysis: reverse osmosis hybrid systems (ED-RO)

Increasing recovery in RO systems requires multiple stages and thus significantly increased capital and operation costs [124]. In the electrical desalination systems such as ED compare to the RO membranes, we cannot achieve to high salt rejection alone [125], and this is very important in energy consumption, however, one of the advantages of ED systems, is operation at higher recovery rate, but and low SEC, by scale formation this process eventually limited [80]. The concept of (ED-RO) hybrid system at first in 1981 by Schmoldt et al. was studied [126]. They proposed the use of ED as a second stage to control permeate quality. However, one of the disadvantages of this system was high energy consumption up to 7.94 kWh/m^3 for SWRO system with a concentration of 45,000 ppm, that was due to some problems such as lack of high-flux and high-selectivity membranes in this process [80, 126]. But, in their studies they showed in a desalination plant with capacity of 1000 m^3/day and feed concentration of lower than 4000 mg/L, the investment cost for ED can be lower than RO. They noted that with the development of the high flux membranes and with high salt-rejection, not only the cost of the RO system could be reduced, hence, with incoming feed with lower TDS concentration, the energy consumption of the ED unit also reduce [126].

In a another study, by Turek et al. [127], in order to assessment SEC and recovery rates, in four different configurations (single-stage standalone RO, NF-SWRO, hybrid ED-RO and NF-SWRO-ED system) for seawater desalination plants were been compared. As can be seen in **Table 8**, the highest recovery (81.1%) was achieved for SWRO-ED, but, at this recovery rate, the SEC was 7.77 kWh/m^3, after that the NFSWRO-ED system had more recovery rate (69.0%) at lower SEC (6.90 kWh/m^3). Although SEC in the SWRO system was much less (2.76 kWh/m^3), but on the other hand, this single-stage RO system operated at a recovery rate of only 43% [80].

In **Table 9**, a comparison of selected ED-RO studies has been presented.

5.2.2 Reverse osmosis: membrane distillation hybrid systems

Several advantages of MD system like as operation at high recovery, high separation efficiency and Low capital cost, has made it alternative candidate for hybrid separation technologies [132, 133]. Over the last few years, a few studies on the hybridization of MD and RO in order to treatment of the concentrate stream from the RO process have been done. For example, in a study by Choi et al., economic feasibility of a RO-MD system for desalination of seawater was assessed. In this study, they found that a RO-MD hybrid system or a MD stand-alone system only when the flux and recovery are greater than that for RO, and or the thermal energy

System	Energy consumption [kWh m^{-3}]	Water recovery [%]
SWRO	2.76	42.6
SWRO-ED	7.77	81.1
NF-SWRO	3.93	41.2
NF-SWRO-ED	6.90	69.0

This article was published in [80], Page 12 and 14, Copyright © 2021 Elsevier B.V (ScienceDirect) (2020).

Table 8.
The SEC and water recovery for SWRO-ED, SWRO, NF-SWRO and NF-SWRO-ED systems [80].

Feed type	Hybridization	Feed TDS (mg/L)	Product TDS (mg/L)	Recovery rate	SEC (kWh/m^3)	Refs.
Brackish water	ED as pretreatment to lower RO feed salinity	2000–4000	50–120	RO alone: 10–20% ED-RO: 50–60%	RO alone: 7.8 ED-RO: 8–10	[128]
Wastewater	ED of RO concentrate	2550–3550	—	RO alone: 75% ED-RO: 95%	—	[129]
Brackish water	ED of RO concentrate; ED product water blended with RO permeate to produce water	3000	300 Hybrid preferred over ED alone only when product TDS requirement is strict	50%	—	[130]
Hypersaline brine	Counterflow ED with RO	120,000	—	Performance at high recoveries is limited by concentration differences	—	[131]

This article was published in [80], Page 12 and 14, Copyright © 2021 Elsevier B.V (ScienceDirect) (2020).

Table 9.
Key parameters from selected ED-RO hybridization studies [80].

that has been supplied for MD, had relatively low cost, can compete with RO system [134]. Although, MD is able to achieve a high water recovery rate of 85%, However, the Energy consumption for RO-MD hybrid systems is still unclear and should be further investigated [80].

5.2.3 Forward osmosis (FO)-RO

Table 10 shows the summary of hybrid FO-RO system for seawater desalination [135].

5.2.4 Nanofiltration (NF)-RO

Using of MF, UF membrane although can be effective for the pretreatment of a SWRO system, but some important parameter such as NOMs, organic matters and dissolved organic matters cannot be fully removed. Since in MF and UF divalent metal ions do not remove, so, the potential of the Scaling cannot be reduced. As we know, in SWRO desalination facility, about 44% of water production costs are related to energy consumption, which is closely related to the salinity of seawater. Hence, in order to pretreatment and effectively reduction of overall salinity (reduce

System	System Detail	FO		RO	Effect	Ref.
		Membrane	Draw solution			
FO-RO	Glucose draw solution (DS) is diluted by seawater at FO and diluted glucose solution is subjected to RO to recover water	—	Glucose	Low pressure reverse osmosis (LPRO)	Low osmotic pressure of glucose, high internal concentration polarization (ICP)	[136]
FO-RO	Secondary waste water is supplied to FO to dilute Red Sea water, which is then subjected to RO	CTA	Red Sea water	LPRO	Energy requirement 50% of SWRO (1.5 kWh/m^3)	[137]
FO-RO-FO	Secondary wastewater is supplied to FO to dilute seawater, which is then subjected to RO to obtain product water. RO brine goes to second FO to be diluted before discharge.	CTA		SW30 2540 Dow Filmtec	Wide range of organic compounds can be removed by FO	[138]
Pressure assisted FO (PAFO)-RO	Wastewater supplied to FO to dilute seawater, which is then subjected to RO				Simulation pressure assisted FO (PAFO) at 6 bar further reduces the water production cost. System operation is stabilized	[139]

Table 10.
Summary of hybrid FO-RO system for seawater desalination [135].

divalent cations) in SWRO system, nanofiltration (NF) can be used [140–142]. In **Table 11**, the summary of the NF-RO hybrid systems is shown. From the view point of the energy consumption, addition of NF pretreatment will increase the energy consumption due to the added pumping energy. However, due to the reduction of salinity in the influent feed solution of RO, the energy consumption decrease [135].

5.2.5 Pressure-retarded osmosis (PRO)-RO

Pressure-retarded osmosis (PRO) is a device to generate power using osmosis.

There are two advantages of coupling SWRO and PRO; (1) enhancement of the power generation in PRO due to the higher osmotic pressure of concentrated brine than seawater, (2) dilution of the concentrated brine before discharging to the ocean In order to combination of RO and PRO there are many different ways, but they can be classified in two groups. First one is transferring the high pressure of DS to the RO feed by using of pressure exchanger and other is generation of electricity with high-pressure DS that spins the turbine. So, with these changes, the specific energy required for water production is reduced (**Figures 11–13**) [159].

There are a number of simulation studies for the RO-PRO hybrid system but only few experimental works have been done using either a small lab-scale equipment or a large demonstration plant, as summarized in **Table 12**, which was made based on the work of Kim et al. [135, 159].

Plant or organization	Pretreatment system	Effect	Refs.
Saline water conversion corporation (SWCC)	Dual and fine sand media filtration (DFSMF)-NF (DFSMF)-NF for RO-multiflush distillation (MFD)	Reduction of total hardness 93%, and TDS 57.7% by NF, MFD operable at distillation temperature of 120°C	[143]
	(DFSMF)-NF	Production of SWRO increased >60% with 30% cost reduction	[144]
Umm Lujj, Saudi Arabia	(DFSMF)-NF	Demonstration plant construction based on the above work	[145]
	NF	Removal of colloidal matters and inorganic scale matters was possible	[146]
	UF-NF	96.3% TOC was removed with 0.06–0.36 mg/L TOC in the filtrate. Gradual membrane fouling was observed	[34]
	NF for RO-MD	Water production cost of 0.92 $/m3 with recovery factor of 76.2%	[147, 148]
	NF-RO-Membrane Crystallization (MCr) NF for RO-MD	It was possible to remove hardness, turbidity, microorganisms, and to reduce chemical and energy consumption. Water production cost was reduced 30%	[149, 150]
Desalination household scale plant (Luna Water 100 GPD)	NF, RO, and NF-RO	Hybrid was the best with rejections of salinity 78.65, TDS 76.52, EC 76.42, Cl 63.95, and Na 70.91%	[151]
Treatment of mine impaired water	Fertilizer drawn FO (FDFONF) is compared with MF-RO and UF-RO	Energy consumption for FDFO-NF was 1.08 kWh/m^3, which is 13.6% less energy than an MF-RO and 21% less than UF-RO	[152]

Table 11.
Summary of NF-RO hybrid system [135].

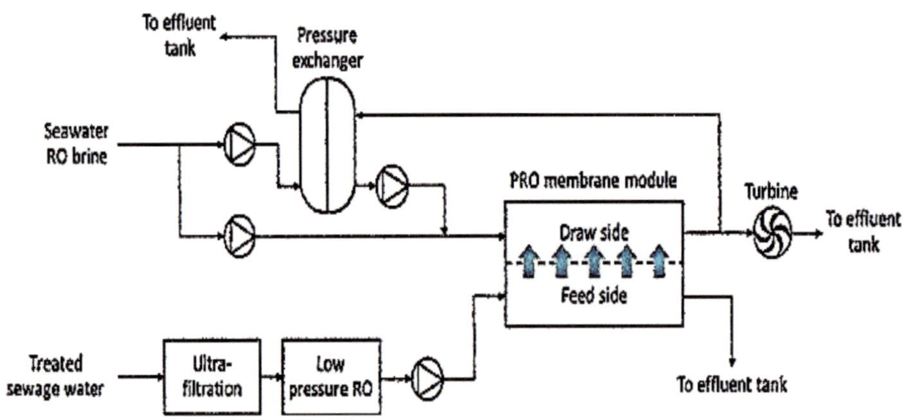

Figure 11.
RO-PRO system in Japan [153].

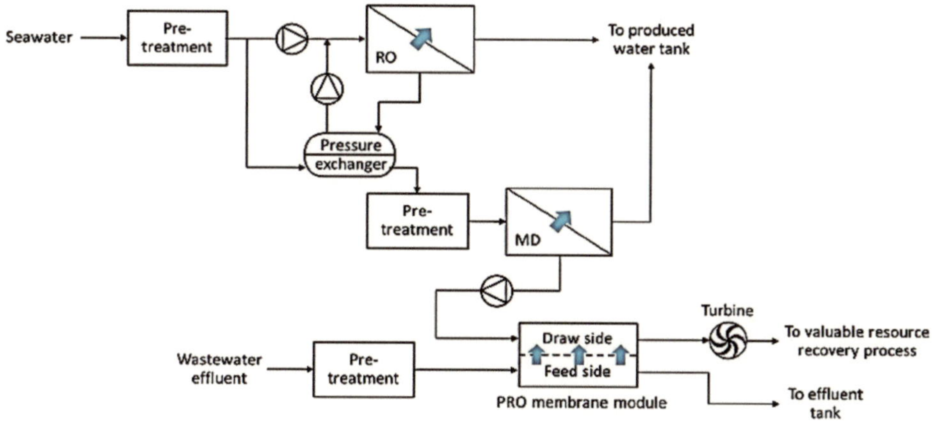

Figure 12.
PRO-MD-RO system in Korea [154].

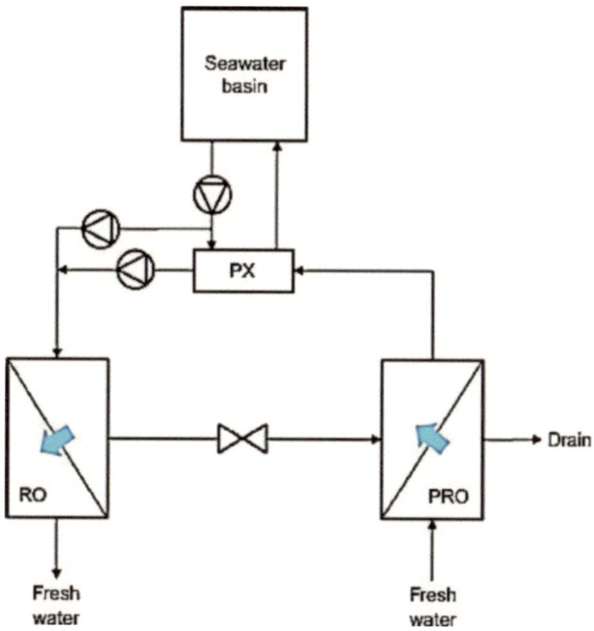

Figure 13.
RO-PRO system of Achilli et al. [80].

6. Conclusion

Considering that consumption of the Energy in hybrid systems, especially for FO-MD, RO-MD and FD-MD processes, due to different operating conditions in many studies are still unclear, we need more research to expand their use in the desalination industry. Research efforts should be directed towards design improvement and evaluation of energy consumption.

Elimination of the restrictions on the use of salinity gradient power technologies and directing them towards commercialization would render hybrid desalination systems more economically and also could use the salinity gradient power as an energy recovery system on their own or with other ERDs in desalination systems as could be used as. In addition to the development of low-cost high power density

System	System detail	PRO		RO	Effect	Refs.
		Membrane	Draw Solution			
RO-PRO	RO brine goes to DS side and pretreated wastewater goes to feed side of PRO	CTA hollow fiber (Toyobo)	RO brine		7.7 W/m^2 was obtained at 2.5 MPa	[153]
RO-MD-PRO	RO brine goes to MD to be further concentrated. MD brine goes to the DS side and pretreated wastewater goes to the feed side of PRO				RO and MD water production capacity of 1000m^3/day and 400 m^3/day, respectively, was achieved with power density of 5 W/m^2	[154]
RO-PRO	RO brine goes to DS, filtrated tap water goes to the feed side of PRO High pressure of DS is transferred to seawater inlet	4040 PRO module (Oasys Water)	RO brine	SW30–2540 (Dow Film Tec)	Power density of 1.1–2.3 W/m^2 was obtained	[155]
RO-PRO	Same as above	CTA membrane (HTI)	RO brine	SW30–4040 (Dow Film Tec)	Simulation based on the experimental data obtained from RO and PRO subsystem. Net specific power consumption for water production is 1.2 kWh/m^3 at 50% RO recovery, 40% less than RO standalone	[156]
RO-PRO					Economic evaluation of RO-PRO hybrid system using model equations	[157]
RO-PRO		10-in hollow Fiber module	RO brine	Toray low pressure RO	13.5 W/m^2 membrane power density. On top of 20% energy reduction by low-pressure RO membrane and RED further 10% energy saving was possible	[158]

Table 12.
Some experimental results of PRO-RO hybrid system [135].

membranes and systems for reverse electrodialysis and pressure retarded osmosis, the implementation and testing of pilot plants would speed up their transition and make them more commercially viable for industrial scale operation with other desalination processes [80].

Acknowledgements

I thank the Desalination publication (Ahmed FE, Hashaikeh R, Hilal N. Hybrid technologies; 2020, Desalination) and (Yang Z, Ma XH, Tang CY; 2018, Desalination) to cultivate the idea of gathering information this book chapter.

Author details

Saeed Pourkarim Nozhdehi
The Superintendent of Rasht Wastewater Treatment Plant Office, Guilan Water and Wastewater Co., Rasht, Iran

*Address all correspondence to: saeedpoorkareem@yahoo.com

References

[1] Elimelech M, Phillip WA. The future of seawater desalination: Energy, technology, and the environment. Science. 2011;**333**:712-717

[2] Hondo H. Life cycle GHG emission analysis of power generation systems: Japanese case. Energy. 2005;**30**: 2042-2056

[3] Roy S, Ragunath S. Emerging membrane technologies for water and energy sustainability: Future prospects, constraints and challenges. Energies. 2018;**11**:2997

[4] Ang WL, Mohammad AW, Hilal N, Leo CP. A review on the applicability of integrated/hybrid membrane processes in water treatment and desalination plants. Desalination. 2015; **363**:2-18

[5] Ali A, Tufa RA, Macedonio F, Curcio E, Drioli E. Membrane technology in renewable-energy-driven desalination. Renewable and Sustainable Energy Reviews. 2018;**81**:1-21

[6] Sanders DF, Smith ZP, Guo R, Robeson LM, McGrath JE, Paul DR, et al. Energy-efficient polymeric gas separation membranes for a sustainable future: A review. Polymer. 2013;**54**: 4729-4761

[7] Yip NY, Tiraferri A, Phillip WA, Schiffman JD, Hoover LA, Kim YC, et al. Thin-film composite pressure retarded osmosis membranes for sustainable power generation from salinity gradients. Environmental Science & Technology. 2011;**45**: 4360-4369

[8] Chu S, Majumdar A. Opportunities and challenges for a sustainable energy future. Nature. 2012;**488**:294-303

[9] Le NL, Nunes SP. Materials and membrane technologies for water and energy sustainability, Sustainable. Materials and Technologies. 2016;**7**:1-28

[10] Cohen-Tanugi D, McGovern RK, Dave SH, Lienhard JH, Grossman JC. Quantifying the potential of ultra-permeable membranes for water desalination. Energy & Environmental Science. 2014;**7**:1134-1141

[11] Yang Z, Ma X-H, Tang CY. Recent development of novel membranes for desalination. Desalination. 2018;**434**: 37-59

[12] Okamoto Y, Lienhard JH. How RO membrane permeability and other performance factors affect process cost and energy use: A review. Desalination. 2019;**470**:114064

[13] Voutchkov N. Energy use for membrane seawater desalination–current status and trends. Desalination. 2018;**431**:2-14

[14] Young M, Esau C. Charting our water future: Economic frameworks to inform decision-making. In: Investing in Water for a Green Economy. London: Routledge; 2015. pp. 67-79

[15] Bartels C, Franks R, Andes K. Operational Performance and Optimization of RO Wastewater Treatment Plants. Singapore: Singapore International Water Week; 2010

[16] Zhu A, Christofides PD, Cohen Y. Effect of thermodynamic restriction on energy cost optimization of RO membrane water desalination. Industrial & Engineering Chemistry Research. 2009;**48**:6010-6021

[17] Gorenflo A, Redondo J, Reverberi F. Basic options and two case studies for retrofitting hollow fiber elements by spiral-wound RO technology. Desalination. 2005;**178**:247-260

[18] MacHarg J, Seacord TF, Sessions B. ADC baseline tests reveal trends in membrane performance. Desalination & Water Reuse. 2008;**18**:30-39

[19] Wilf M. Effect of new generation of low pressure, high salt rejection membranes on power consumption of RO systems. In: Proceedings of AWWA Membrane Technology Conference. New Orleans; 1997. pp. 663-679

[20] Franks R, Bartels CR, Andes K, Patel M, Young T. Implementing energy saving RO technology in large scale wastewater treatment plants. In: Proceedings of the International Desalination and Water Reuse Conference, Las Palmas, Spain, Citeseer. 2007

[21] Zhu A, Christofides PD, Cohen Y. On RO membrane and energy costs and associated incentives for future enhancements of membrane permeability. Journal of Membrane Science. 2009;**344**:1-5

[22] Shrivastava A, Rosenberg S, Peery M. Energy efficiency breakdown of reverse osmosis and its implications on future innovation roadmap for desalination. Desalination. 2015;**368**: 181-192

[23] McGovern RK. On the asymptotic flux of ultrapermeable seawater reverse osmosis membranes due to concentration polarisation. Journal of Membrane Science. 2016;**520**:560-565

[24] Werber JR, Deshmukh A, Elimelech M. The critical need for increased selectivity, not increased water permeability, for desalination membranes. Environmental Science & Technology Letters. 2016;**3**:112-120

[25] Mazlan NM, Peshev D, Livingston AG. Energy consumption for desalination—A comparison of forward osmosis with reverse osmosis, and the potential for perfect membranes. Desalination. 2016;**377**:138-151

[26] Shi B, Marchetti P, Peshev D, Zhang S, Livingston AG. Will ultra-high permeance membranes lead to ultra-efficient processes? Challenges for molecular separations in liquid systems. Journal of Membrane Science. 2017;**525**: 35-47

[27] Wei QJ, McGovern RK. Saving energy with an optimized two-stage reverse osmosis system. Environmental Science: Water Research & Technology. 2017;**3**:659-670

[28] Karabelas A, Koutsou C, Kostoglou M, Sioutopoulos D. Analysis of specific energy consumption in reverse osmosis desalination processes. Desalination. 2018;**431**:15-21

[29] Busch M, Mickols W. Reducing energy consumption in seawater desalination. Desalination. 2004;**165**: 299-312

[30] A.W.W. Association. Reverse Osmosis and Nanofiltration: Manual of Water Supply Practices (M46). Denver, CO: American Water Works Association; 2007

[31] Rodriguez-Calvo A, Silva-Castro GA, Osorio F, Gonzalez-Lopez J, Calvo C. Reverse osmosis seawater desalination: Current status of membrane systems. Desalination and Water Treatment. 2015;**56**:849-861

[32] Rashid M, Ralph SF. Carbon nanotube membranes: Synthesis, properties, and future filtration applications. Nanomaterials. 2017;7:99

[33] Goh P, Matsuura T, Ismail A, Hilal N. Recent trends in membranes and membrane processes for desalination. Desalination. 2016;**391**:43-60

[34] Fontananova E, Di Profio G, Artusa F, Drioli E. Polymeric homogeneous

composite membranes for separations in organic solvents. Journal of Applied Polymer Science. 2013;**129**:1653-1659

[35] Gorgojo P, Karan S, Wong HC, Jimenez-Solomon MF, Cabral JT, Livingston AG. Ultrathin polymer films with intrinsic microporosity: Anomalous solvent permeation and high flux membranes. Advanced Functional Materials. 2014;**24**:4729-4737

[36] Greenlee LF, Lawler DF, Freeman BD, Marrot B, Moulin P. Reverse osmosis desalination: Water sources, technology, and today's challenges. Water Research. 2009;**43**:2317-2348

[37] Khedr MG. Development of reverse osmosis desalination membranes composition and configuration: Future prospects. Desalination. 2003;**153**: 295-304

[38] Park HB, Freeman BD, Zhang ZB, Sankir M, McGrath JE. Highly chlorine-tolerant polymers for desalination. Angewandte Chemie International Edition. 2008;**47**:6019-6024

[39] Werber JR, Osuji CO, Elimelech M. Materials for next-generation desalination and water purification membranes. Nature Reviews Materials. 2016;**1**:1-15

[40] Fane A, Tang C, Wang R. Membrane Technology for Water: Microfiltration, Ultrafiltration, Nanofiltration, and Reverse Osmosis. Nanyang Technological University, Singapore: Elsevier; 2011. pp. 301-335

[41] Kumar M, Grzelakowski M, Zilles J, Clark M, Meier W. Highly permeable polymeric membranes based on the incorporation of the functional water channel protein Aquaporin Z. Proceedings of the National Academy of Sciences. 2007;**104**:20719-20724

[42] Shen Y-X, Saboe PO, Sines IT, Erbakan M, Kumar M. Biomimetic membranes: A review. Journal of Membrane Science. 2014;**454**:359-381

[43] Hinds BJ, Chopra N, Rantell T, Andrews R, Gavalas V, Bachas LG. Aligned multiwalled carbon nanotube membranes. Science. 2004;**303**:62-65

[44] Nair R, Wu H, Jayaram P, Grigorieva I, Geim A. Unimpeded permeation of water through helium-leak–tight graphene-based membranes. Science. 2012;**335**:442-444

[45] Jeong B-H, Hoek EM, Yan Y, Subramani A, Huang X, Hurwitz G, et al. Interfacial polymerization of thin film nanocomposites: A new concept for reverse osmosis membranes. Journal of Membrane Science. 2007;**294**:1-7

[46] Yin J, Kim E-S, Yang J, Deng B. Fabrication of a novel thin-film nanocomposite (TFN) membrane containing MCM-41 silica nanoparticles (NPs) for water purification. Journal of Membrane Science. 2012;**423**:238-246

[47] Surwade SP, Smirnov SN, Vlassiouk IV, Unocic RR, Veith GM, Dai S, et al. Water desalination using nanoporous single-layer graphene. Nature Nanotechnology. 2015;**10**:459-464

[48] Cohen-Tanugi D, Grossman JC. Nanoporous graphene as a reverse osmosis membrane: Recent insights from theory and simulation. Desalination. 2015;**366**:59-70

[49] Song X, Wang L, Tang CY, Wang Z, Gao C. Fabrication of carbon nanotubes incorporated double-skinned thin film nanocomposite membranes for enhanced separation performance and antifouling capability in forward osmosis process. Desalination. 2015;**369**: 1-9

[50] Xue S-M, Xu Z-L, Tang Y-J, Ji C-H. Polypiperazine-amide nanofiltration membrane modified by different functionalized multiwalled carbon

nanotubes (MWCNTs). ACS Applied Materials & Interfaces. 2016;**8**: 19135-19144

[51] Wang J, Zhang P, Liang B, Liu Y, Xu T, Wang L, et al. Graphene oxide as an effective barrier on a porous nanofibrous membrane for water treatment. ACS Applied Materials & Interfaces. 2016;**8**:6211-6218

[52] Fujioka T, Oshima N, Suzuki R, Price WE, Nghiem LD. Probing the internal structure of reverse osmosis membranes by positron annihilation spectroscopy: Gaining more insight into the transport of water and small solutes. Journal of Membrane Science. 2015;**486**: 106-118

[53] Freger V. Nanoscale heterogeneity of polyamide membranes formed by interfacial polymerization. Langmuir. 2003;**19**:4791-4797

[54] Jung JS, Preston GM, Smith BL, Guggino WB, Agre P. Molecular structure of the water channel through aquaporin CHIP. The hourglass model. Journal of Biological Chemistry. 1994; **269**:14648-14654

[55] Agre P. Aquaporin water channels (Nobel lecture). Angewandte Chemie International Edition. 2004;**43**: 4278-4290

[56] Holt JK, Park HG, Wang Y, Stadermann M, Artyukhin AB, Grigoropoulos CP, et al. Fast mass transport through sub-2-nanometer carbon nanotubes. Science. 2006;**312**: 1034-1037

[57] Abraham J, Vasu KS, Williams CD, Gopinadhan K, Su Y, Cherian CT, et al. Tunable sieving of ions using graphene oxide membranes. Nature Nanotechnology. 2017;**12**:546-550

[58] Kang G-D, Gao C-J, Chen W-D, Jie X-M, Cao Y-M, Yuan Q. Study on hypochlorite degradation of aromatic polyamide reverse osmosis membrane. Journal of Membrane Science. 2007;**300**: 165-171

[59] Yin J, Zhu G, Deng B. Multi-walled carbon nanotubes (MWNTs)/ polysulfone (PSU) mixed matrix hollow fiber membranes for enhanced water treatment. Journal of Membrane Science. 2013;**437**:237-248

[60] Liu S, Zeng TH, Hofmann M, Burcombe E, Wei J, Jiang R, et al. Antibacterial activity of graphite, graphite oxide, graphene oxide, and reduced graphene oxide: Membrane and oxidative stress. ACS Nano. 2011;**5**: 6971-6980

[61] Li X, Wang R, Tang C, Vararattanavech A, Zhao Y, Torres J, et al. Preparation of supported lipid membranes for aquaporin Z incorporation. Colloids and Surfaces B: Biointerfaces. 2012;**94**:333-340

[62] Zhong PS, Chung T-S, Jeyaseelan K, Armugam A. Aquaporin-embedded biomimetic membranes for nanofiltration. Journal of Membrane Science. 2012;**407**:27-33

[63] Duong PH, Chung T-S, Jeyaseelan K, Armugam A, Chen Z, Yang J, et al. Planar biomimetic aquaporin-incorporated triblock copolymer membranes on porous alumina supports for nanofiltration. Journal of Membrane Science. 2012;**409**:34-43

[64] Wang M, Wang Z, Wang X, Wang S, Ding W, Gao C. Layer-by-layer assembly of aquaporin Z-incorporated biomimetic membranes for water purification. Environmental Science & Technology. 2015;**49**:3761-3768

[65] Wang H, Chung TS, Tong YW, Jeyaseelan K, Armugam A, Chen Z, et al. Highly permeable and selective pore-spanning biomimetic membrane embedded with aquaporin Z. Small. 2012;**8**:1185-1190

[66] Ding W, Cai J, Yu Z, Wang Q, Xu Z, Wang Z, et al. Fabrication of an aquaporin-based forward osmosis membrane through covalent bonding of a lipid bilayer to a microporous support. Journal of Materials Chemistry A. 2015; **3**:20118-20126

[67] Zhao Y, Qiu C, Li X, Vararattanavech A, Shen W, Torres J, et al. Synthesis of robust and high-performance aquaporin-based biomimetic membranes by interfacial polymerization-membrane preparation and RO performance characterization. Journal of Membrane Science. 2012;**423**: 422-428

[68] Li X, Chou S, Wang R, Shi L, Fang W, Chaitra G, et al. Nature gives the best solution for desalination: Aquaporin-based hollow fiber composite membrane with superior performance. Journal of Membrane Science. 2015;**494**:68-77

[69] Qi S, Wang R, Chaitra GKM, Torres J, Hu X, Fane AG. Aquaporin-based biomimetic reverse osmosis membranes: Stability and long term performance. Journal of Membrane Science. 2016;**508**: 94-103

[70] Sun G, Chung T-S, Jeyaseelan K, Armugam A. A layer-by-layer self-assembly approach to developing an aquaporin-embedded mixed matrix membrane. RSC Advances. 2013;**3**: 473-481

[71] Li X, Wang R, Wicaksana F, Tang C, Torres J, Fane AG. Preparation of high performance nanofiltration (NF) membranes incorporated with aquaporin Z. Journal of Membrane Science. 2014;**450**:181-188

[72] Xie W, He F, Wang B, Chung T-S, Jeyaseelan K, Armugam A, et al. An aquaporin-based vesicle-embedded polymeric membrane for low energy water filtration. Journal of Materials Chemistry A. 2013;**1**:7592-7600

[73] Wang HL, Chung T-S, Tong YW, Jeyaseelan K, Armugam A, Duong HHP, et al. Mechanically robust and highly permeable AquaporinZ biomimetic membranes. Journal of Membrane Science. 2013;**434**:130-136

[74] Sun G, Chung T-S, Chen N, Lu X, Zhao Q. Highly permeable aquaporin-embedded biomimetic membranes featuring a magnetic-aided approach. RSC Advances. 2013;**3**:9178-9184

[75] Hummer G, Rasaiah J, Noworyta J. Nanoscale hydrodynamics: Enhanced flow in carbon nanotubes. Nature. 2001; **414**:188

[76] Manawi Y, Kochkodan V, Hussein MA, Khaleel MA, Khraisheh M, Hilal N. Can carbon-based nanomaterials revolutionize membrane fabrication for water treatment and desalination? Desalination. 2016;**391**:69-88

[77] Das R, Ali ME, Abd Hamid SB, Ramakrishna S, Chowdhury ZZ. Carbon nanotube membranes for water purification: A bright future in water desalination. Desalination. 2014;**336**: 97-109

[78] Hegab HM, Zou L. Graphene oxide-assisted membranes: Fabrication and potential applications in desalination and water purification. Journal of Membrane Science. 2015;**484**:95-106

[79] Wang K, Abdalla AA, Khaleel MA, Hilal N, Khraisheh MK. Mechanical properties of water desalination and wastewater treatment membranes. Desalination. 2017;**401**:190-205

[80] Ahmed FE, Hashaikeh R, Hilal N. Hybrid technologies: The future of energy efficient desalination–A review. Desalination. 2020;**495**:114659

[81] Ghaffour N, Missimer TM, Amy GL. Technical review and evaluation of the economics of water desalination: Current and future challenges for better

water supply sustainability. Desalination. 2013;**309**:197-207

[82] Micale G, Rizzuti L, Cipollina A. Seawater Desalination: Conventional and Renewable Energy Processes. Berlin Heidelberg: Springer; 2009

[83] Gude VG. Desalination and sustainability—An appraisal and current perspective. Water Research. 2016;**89**: 87-106

[84] Al-Mutaz IS, Al-Namlah AM. Characteristics of dual purpose MSF desalination plants. Desalination. 2004; **166**:287-294

[85] Darwish M, Al-Najem NM. Energy consumption by multi-stage flash and reverse osmosis desalters. Applied Thermal Engineering. 2000;**20**:399-416

[86] El-Naser H. Management of Scarce Water Resources: A Middle Eastern Experience. UK: WIT Press; 2009

[87] Chua HT, Rahimi B. Low Grade Heat Driven Multi-effect Distillation and Desalination. Netherlands: Elsevier; 2017

[88] Al-Shammiri M, Safar M. Multi-effect distillation plants: State of the art. Desalination. 1999;**126**:45-59

[89] Sorribas S, Gorgojo P, Téllez C, Coronas J, Livingston AG. High flux thin film nanocomposite membranes based on metal–organic frameworks for organic solvent nanofiltration. Journal of the American Chemical Society. 2013; **135**:15201-15208

[90] Wang X, Christ A, Regenauer-Lieb K, Hooman K, Chua HT. Low grade heat driven multi-effect distillation technology. International Journal of Heat and Mass Transfer. 2011;**54**: 5497-5503

[91] Ghaffour N. The challenge of capacity-building strategies and perspectives for desalination for sustainable water use in MENA. Desalination and Water Treatment. 2009;**5**:48-53

[92] Sajtar ET, Bagley DM. Electrodialysis reversal: Process and cost approximations for treating coal-bed methane waters. Desalination and Water Treatment. 2009;**2**:284-294

[93] Strathmann H. Overview of ion-exchange membrane processes. Ion-Exchange Membrane Separation Processes. 2004;**9**:1-22

[94] Leitz F, Accomazzo M, McRae W. High temperature electrodialysis. Desalination. 1974;**14**:33-41

[95] Korngold E. Electrodialysis unit: Optimization and calculation of energy requirement. Desalination. 1982;**40**: 171-179

[96] Ferreira JZ, Bernardes AM, Rodrigues MAS. Electrodialysis and Water Reuse: Novel Approaches. Berlin, Heidelberg: Springer; 2014

[97] Benneker AM, Rijnaarts T, Lammertink RG, Wood JA. Effect of temperature gradients in (reverse) electrodialysis in the Ohmic regime. Journal of Membrane Science. 2018;**548**: 421-428

[98] Rizzuti L, Ettouney HM, Cipollina A. Solar Desalination for the 21st Century: A Review of Modern Technologies and Researches on Desalination Coupled to Renewable Energies. Berlin, Heidelberg: Springer Science & Business Media; 2007

[99] Chehayeb KM, Nayar KG. On the merits of using multi-stage and counterflow electrodialysis for reduced energy consumption. Desalination. 2018;**439**:1-16

[100] Xu T, Huang C. Electrodialysis-based separation technologies: A critical

review. AICHE Journal. 2008;**54**: 3147-3159

[101] Warsinger DM, Servi A, Connors GB, Mavukkandy MO, Arafat HA, Gleason KK. Reversing membrane wetting in membrane distillation: Comparing dryout to backwashing with pressurized air. Environmental Science: Water Research & Technology. 2017;**3**: 930-939

[102] Drioli E, Ali A, Macedonio F. Membrane distillation: Recent developments and perspectives. Desalination. 2015;**356**:56-84

[103] Amy G, Ghaffour N, Li Z, Francis L, Linares RV, Missimer T, et al. Membrane-based seawater desalination: Present and future prospects. Desalination. 2017;**401**:16-21

[104] Zakrzewska-Trznadel G, Harasimowicz M, Chmielewski AG. Concentration of radioactive components in liquid low-level radioactive waste by membrane distillation. Journal of Membrane Science. 1999;**163**:257-264

[105] Koschikowski J, Wieghaus M, Rommel M. Solar thermal-driven desalination plants based on membrane distillation. Desalination. 2003;**156**: 295-304

[106] Bouguecha S, Hamrouni B, Dhahbi M. Small scale desalination pilots powered by renewable energy sources: Case studies. Desalination. 2005;**183**: 151-165

[107] Banat F, Jwaied N, Rommel M, Koschikowski J, Wieghaus M. Performance evaluation of the "large SMADES" autonomous desalination solar-driven membrane distillation plant in Aqaba, Jordan. Desalination. 2007; **217**:17-28

[108] Macedonio F, Curcio E, Drioli E. Integrated membrane systems for seawater desalination: Energetic and exergetic analysis, economic evaluation, experimental study. Desalination. 2007; **203**:260-276

[109] Criscuoli A, Carnevale MC, Drioli E. Evaluation of energy requirements in membrane distillation. Chemical Engineering and Processing: Process Intensification. 2008;**47**:1098-1105

[110] Wang X, Zhang L, Yang H, Chen H. Feasibility research of potable water production via solar-heated hollow fiber membrane distillation system. Desalination. 2009;**247**:403-411

[111] Duong HC, Cooper P, Nelemans B, Cath TY, Nghiem LD. Evaluating energy consumption of air gap membrane distillation for seawater desalination at pilot scale level. Separation and Purification Technology. 2016;**166**:55-62

[112] Criscuoli A, Carnevale M, Drioli E. Modeling the performance of flat and capillary membrane modules in vacuum membrane distillation. Journal of Membrane Science. 2013;**447**:369-375

[113] Guan G, Yang X, Wang R, Field R, Fane AG. Evaluation of hollow fiber-based direct contact and vacuum membrane distillation systems using aspen process simulation. Journal of Membrane Science. 2014;**464**:127-139

[114] Dow N, Gray S, Zhang J, Ostarcevic E, Liubinas A, Atherton P, et al. Pilot trial of membrane distillation driven by low grade waste heat: Membrane fouling and energy assessment. Desalination. 2016;**391**: 30-42

[115] Ali MI, Summers EK, Arafat HA. Effects of membrane properties on water production cost in small scale membrane distillation systems. Desalination. 2012;**306**:60-71

[116] Drioli E, Criscuoli A, Curcio E. Membrane Contactors: Fundamentals,

Applications and Potentialities. Netherlands: Elsevier; 2011

[117] Summers EK, Arafat HA. Energy efficiency comparison of single-stage membrane distillation (MD) desalination cycles in different configurations. Desalination. 2012;**290**: 54-66

[118] Jantaporn W, Ali A, Aimar P. Specific energy requirement of direct contact membrane distillation. Chemical Engineering Research and Design. 2017; **128**:15-26

[119] Feher J. Osmosis and osmotic pressure. Quantitative Human Physiology. 2012;**10**:141-152

[120] Eyvaz M, Arslan S, İmer D, Yüksel E, Koyuncu İ. Forward osmosis membranes–A review: Part I. In: Osmotically Driven Membrane Processes-Approach, Development and Current Status. London, UK: IntechOpen; 2018. pp. 11-40

[121] McGovern RK. On the potential of forward osmosis to energetically outperform reverse osmosis desalination. Journal of Membrane Science. 2014;**469**:245-250

[122] Awad AM, Jalab R, Minier-Matar J, Adham S, Nasser MS, Judd S. The status of forward osmosis technology implementation. Desalination. 2019; **461**:10-21

[123] Moon AS, Lee M. Energy consumption in forward osmosis-desalination compared to other desalination techniques. World Academy of Science Engineering and Technology. 2012;**65**:537-539

[124] Stover RL. Industrial and brackish water treatment with closed circuit reverse osmosis. Desalination and Water Treatment. 2013;**51**: 1124-1130

[125] Kim B, Kwak R, Kwon HJ, Kim M, Al-Anzi B, Lim G, et al. Purification of high salinity brine by multi-stage ion concentration polarization desalination. Scientific Reports. 2016;**6**:1-12

[126] Doornbusch G, Tedesco M, Post J, Borneman Z, Nijmeijer K. Experimental investigation of multistage electrodialysis for seawater desalination. Desalination. 2019;**464**:105-114

[127] Turek M, Mitko K, Laskowska E, Chorążewska M, Piotrowski K, Jakóbik-Kolon A, et al. Energy consumption and gypsum scaling assessment in a hybrid nanofiltration-reverse osmosis-electrodialysis system. Chemical Engineering & Technology. 2018;**41**: 392-400

[128] Thampy S, Desale GR, Shahi VK, Makwana BS, Ghosh PK. Development of hybrid electrodialysis-reverse osmosis domestic desalination unit for high recovery of product water. Desalination. 2011;**282**:104-108

[129] Zhang Y, Ghyselbrecht K, Meesschaert B, Pinoy L, Van der Bruggen B. Electrodialysis on RO concentrate to improve water recovery in wastewater reclamation. Journal of Membrane Science. 2011;**378**: 101-110

[130] McGovern RK, Zubair SM. The benefits of hybridising electrodialysis with reverse osmosis. Journal of Membrane Science. 2014;**469**:326-335

[131] McGovern RK, Zubair SM, Lienhard V J. Hybrid electrodialysis reverse osmosis system design and its optimization for treatment of highly saline brines. IDA Journal of Desalination and Water Reuse. 2014;**6**: 15-23

[132] Wang P, Chung T-S. Recent advances in membrane distillation processes: Membrane development, configuration design and application

exploring. Journal of Membrane Science. 2015;**474**:39-56

[133] Eumine Suk D, Matsuura T. Membrane-based hybrid processes: A review. Separation Science and Technology. 2006;**41**:595-626

[134] Choi Y-J, Lee S, Koo J, Kim S-H. Evaluation of economic feasibility of reverse osmosis and membrane distillation hybrid system for desalination. Desalination and Water Treatment. 2016;**57**:24662-24673

[135] Ismail F, Khulbe KC, Matsuura T. Reverse Osmosis. Netherlands: Elsevier; 2018

[136] Volkov AV, Parashchuk VV, Stamatialis DF, Khotimsky VS, Volkov VV, Wessling M. High permeable PTMSP/PAN composite membranes for solvent nanofiltration. Journal of Membrane Science. 2009;**333**:88-93

[137] Tsar'kov S, Malakhov A, Litvinova E, Volkov A. Nanofiltration of dye solutions through membranes based on poly (trimethylsilylpropyne). Petroleum Chemistry. 2013;**53**:537-545

[138] Li X, Vandezande P, Vankelecom IF. Polypyrrole modified solvent resistant nanofiltration membranes. Journal of Membrane Science. 2008;**320**: 143-150

[139] Fritsch D, Merten P, Heinrich K, Lazar M, Priske M. High performance organic solvent nanofiltration membranes: Development and thorough testing of thin film composite membranes made of polymers of intrinsic microporosity (PIMs). Journal of Membrane Science. 2012;**401**:222-231

[140] da Silva Burgal J, Peeva L, Marchetti P, Livingston A. Controlling molecular weight cut-off of PEEK nanofiltration membranes using a drying method. Journal of Membrane Science. 2015;**493**:524-538

[141] Vanherck K, Cano-Odena A, Koeckelberghs G, Dedroog T, Vankelecom I. A simplified diamine crosslinking method for PI nanofiltration membranes. Journal of Membrane Science. 2010;**353**: 135-143

[142] Dutczak S, Cuperus F, Wessling M, Stamatialis D. New crosslinking method of polyamide–imide membranes for potential application in harsh polar aprotic solvents. Separation and Purification Technology. 2013;**102**: 142-146

[143] Huang J-H, Zhou C-F, Zeng G-M, Li X, Niu J, Huang H-J, et al. Micellar-enhanced ultrafiltration of methylene blue from dye wastewater via a polysulfone hollow fiber membrane. Journal of Membrane Science. 2010;**365**: 138-144

[144] Strużyńska-Piron I, Loccufier J, Vanmaele L, Vankelecom IF. Synthesis of solvent stable polymeric membranes via UV depth-curing. Chemical Communications. 2013;**49**: 11494-11496

[145] Strużyńska-Piron I, Loccufier J, Vanmaele L, Vankelecom IF. Parameter study on the preparation of UV depth-cured chemically resistant polysulfone-based membranes. Macromolecular Chemistry and Physics. 2014;**215**: 614-623

[146] Strużyńska-Piron I, Bilad MR, Loccufier J, Vanmaele L, Vankelecom IF. Influence of UV curing on morphology and performance of polysulfone membranes containing acrylates. Journal of Membrane Science. 2014;**462**:17-27

[147] Ohya H, Okazaki I, Aihara M, Tanisho S, Negishi Y. Study on molecular weight cut-off performance of asymmetric aromatic polyimide membrane. Journal of Membrane Science. 1997;**123**:143-147

[148] Valtcheva IB, Kumbharkar SC, Kim JF, Bhole Y, Livingston AG. Beyond polyimide: Crosslinked polybenzimidazole membranes for organic solvent nanofiltration (OSN) in harsh environments. Journal of Membrane Science. 2014;**457**:62-72

[149] Xing DY, Chan SY, Chung T-S. The ionic liquid [EMIM] OAc as a solvent to fabricate stable polybenzimidazole membranes for organic solvent nanofiltration. Green Chemistry. 2014; **16**:1383-1392

[150] Kim JF, Gaffney PR, Valtcheva IB, Williams G, Buswell AM, Anson MS, et al. Organic solvent nanofiltration (OSN): A new technology platform for liquid-phase oligonucleotide synthesis (LPOS). Organic Process Research & Development. 2016;**20**:1439-1452

[151] Xu YC, Cheng XQ, Long J, Shao L. A novel monoamine modification strategy toward high-performance organic solvent nanofiltration (OSN) membrane for sustainable molecular separations. Journal of Membrane Science. 2016;**497**:77-89

[152] Solomon MFJ, Bhole Y, Livingston AG. High flux membranes for organic solvent nanofiltration (OSN)—Interfacial polymerization with solvent activation. Journal of Membrane Science. 2012;**423**:371-382

[153] Huang L, Chen J, Gao T, Zhang M, Li Y, Dai L, et al. Reduced graphene oxide membranes for ultrafast organic solvent nanofiltration. Advanced Materials. 2016;**28**:8669-8674

[154] Soroko I, Livingston A. Impact of TiO_2 nanoparticles on morphology and performance of crosslinked polyimide organic solvent nanofiltration (OSN) membranes. Journal of Membrane Science. 2009;**343**:189-198

[155] Vanherck K, Hermans S, Verbiest T, Vankelecom I. Using the photothermal effect to improve membrane separations via localized heating. Journal of Materials Chemistry. 2011;**21**:6079-6087

[156] Li Y, Verbiest T, Vankelecom I. Improving the flux of PDMS membranes via localized heating through incorporation of gold nanoparticles. Journal of Membrane Science. 2013;**428**:63-69

[157] Vanherck K, Vankelecom I, Verbiest T. Improving fluxes of polyimide membranes containing gold nanoparticles by photothermal heating. Journal of Membrane Science. 2011;**373**:5-13

[158] Campbell J, Székely G, Davies R, Braddock DC, Livingston AG. Fabrication of hybrid polymer/metal organic framework membranes: Mixed matrix membranes versus in situ growth. Journal of Materials Chemistry A. 2014;**2**:9260-9271

[159] Gevers LE, Vankelecom IF, Jacobs PA. Solvent-resistant nanofiltration with filled polydimethylsiloxane (PDMS) membranes. Journal of Membrane Science. 2006;**278**:199-204

Section 2

Membrane and Brine Management

Chapter 2

Desalination Membrane Management

Thallam Lakshmi Prasad

Abstract

With growing market of membrane technologies the disposal of these spent modules going to be serious issue especially for water industry. Review of status of technologies is briefly highlighted. Keeping this in mind, the various schemes/protocols can be planned and accordingly exploratory studies have been initiated using AOP based primary techniques such as hydro thermal processes. This chapter presents both the open literature and experimental studies related to spent desalination membranes.

Keywords: spent membranes, recycling, low temperature AOP processes, irradiation, high temperature processes

1. Introduction

To meet the increasing demand for fresh water under growing environmental awareness and constraints, necessity of desalination techniques are being felt strongly. Potable water by desalination can be produced either by thermal process or membrane processes. Membrane application is an emerging area of interest, as membrane processes operate at ambient temperature and offer one step separation for dissolved constituents on molecular level. Reverse Osmosis (RO) is long established as a large scale industrial membrane process. According to International Desalination Association (IDA) report, the cumulative global installed capacity is now 92.5 million m^3/day with 19,372 Reverse Osmosis (RO) plants around the world. By 30th June 2018, this number has increased to 20,000. The large desalination market in recent years has resulted in increased waste generation associated with this technology, which has led to the disposal of more than 840,000 End-of-Life (EoL) membranes (>14,000 tonne/year) every year worldwide. Literature shows that these membranes can be repaired in order to reuse them for secondary and tertiary purposes. **Figure 1** shows Installed and projected desalination capacity including mostly applied process. Wide variety of polymers such as cellulose acetate; poly acrylonitrile; polyamide and polysulfones are used for RO modules and ultra filtration processes [1].

In most of the cases membranes are deployed in spiral configuration. Depending on the process and operating conditions, these modules have design life and needs to be replaced after 3 to 5 years. This generates lot of spent RO module as waste. With growing market of membrane technologies the disposal of these spent modules going to be serious issues. Review of status of technologies is briefly highlighted. Keeping this in mind, the various schemes/protocols [2] can be planned and accordingly exploratory studies have been initiated on primary techniques which are based on hydro thermal processes.

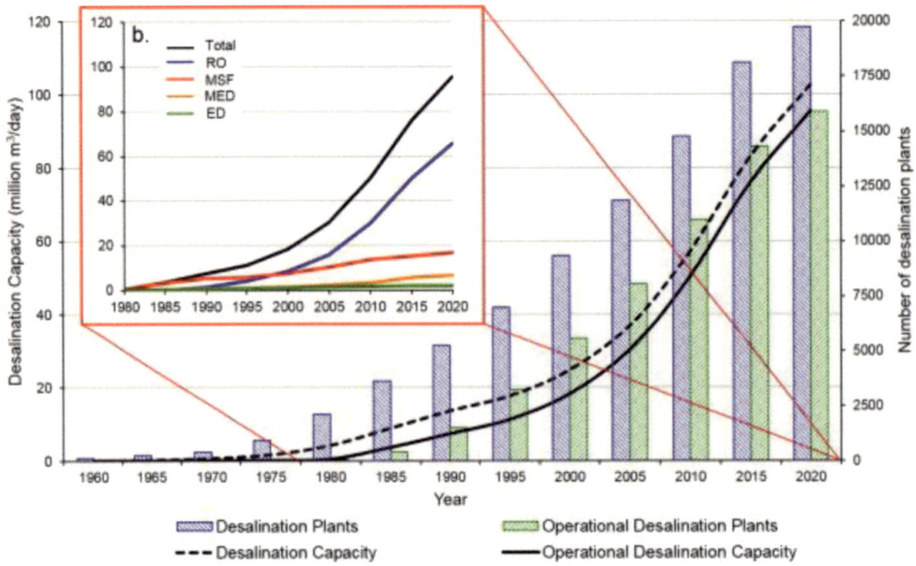

Figure 1.
Installed and projected desalination capacity including mostly applied process.

Hence eco-friendly disposal of spent membranes is an important issue for desalination industry. In our initial approach, we have carried out lab scale studies to study the various hydrothermal process techniques on mineralisation of polyamide thin film composite membrane and as well as poly sulfone membrane. The low temperature AOP processes requires mild chemical duty conditions and in turn helps in bringing down the capital costs of waste treatment plants.

2. Status of processes technologies for solid forms of wastes

The waste treatment status in India is typically as shown in **Figure 2**. The spent desalination membranes being in solid form of waste, the present section of the chapter reviews the status of technologies currently deployed.

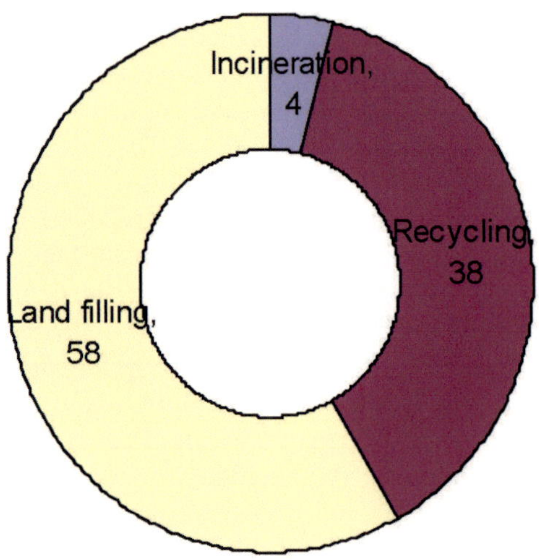

Figure 2.
Waste treatment status in India.

The salient features of some of the existing industrial practices in nuclear industry, as well as other industries are presented in **Table 1** below.

In recent times, several Hydro Thermal Processes (HTPs) have emerged as eco-friendly alternatives to incineration. Some of these processes are Wet Oxidation; Photo Oxidation and Wet Air Oxidation. The HTPs help in complete mineralisation of the organic wastes and forms CO_2 and H_2O. Sources for oxidising the species are air, oxygen, ozone and H_2O_2. The basic chemical reaction in HTPs is

$$\text{Catalyst Organic substrate} + \text{Oxidant} ========\rightarrow CO_2 + H_2O + \text{Mineral acid / salt}$$

Sl. No.	Location	Technical data	Remarks
A)	Immobalisation in cement		
1	Ringhals Nuclear power station Sweden (1975)	i. Mixing of dewatered resin with cement (14% w/w) in prefabricated mould ii. 2000 moulds generated	Sp. activity of resins 3.7x10e2 to 3.7x10e4 G.Bq/cu.mt
2	Federal Republic of Germany (1981)	i. 400 l steel drum mixer ii. 2–3 drums/hr	Do
B)	Immobalisation in Bitumen		
3	Barseback bituminisation unit, Sweden (1975)	i. Feed rate of thin film evaporator is 90 kg/hr. for slurry and 33 kg/hr. for bitumen ii. Op. Temperature is 160°C iii. Capacity 2 drums/day	
4	Finland (1979)	i. Drying in conical steel dryer heated by steam and then mixed with bitumen at 50% w/w ii. Op. Temperature 135°C iii. Capacity 100L/hr	
5	Switzerland (1978)	Extruder of 120 L/hr	
6	Chalk river, Canada (1976)	Thin film evaporator of 100L/hr	
C)	Immobalisation in polymers		
7	Chooz immobalisation unit, France (1981)	i. Polyester in 200 L drum mixing ii. 1–2 drums/day iii. Ambient temperature	Sp. activity of resin 1.85x10e4 G.Bq/cu.mt
8	Fama mobile plant, FRG (1976)	i. Polystyrene in 110 L drum mixing ii. 1–2 cu.mt/day iii. Ambient temperature	
9	Mobile plant, USA(1982)	Modified vinyl ester mixing in container of 4.5 cu.mt	
10	Narora, India (1996)	Polyester in modified drum mixing after dewatering	
D)	Incineration of spent resin		
11	Nine mile island, USA	i. Fluid bed incineration of capacity 12 kg/hr. ii. Off-gas treatment by scrubber, demister, HEPA, iodine absorber	

Sl. No.	Location	Technical data	Remarks
E)	Green chemistry of PET bottles		
	IBM Almaden Research Center, San Jose, CA, USA	i. PET organocatalytic depolymerisation for chemical recycling. ii. Heating at 190°C with catalyst and ethylene glycol iii. Reactants and catalysts can be recycled many times	Research with KACST and Stanford university
F)	Nylon-6 depolymerisation		
	Shangai University China	i. Water medium at temperatures of 553 K to 603 K and pressures of 6.4 to 12.8 MPa ii. Reaction activation energy is 77.38 KJ/mol	Research

Table 1.
The features of the existing industrial practices.

Wet Oxidation is a mineralisation process employing powerful oxidants such as Hydrogen peroxide and Ozone [3]. The reaction occurs in aqueous medium at a maximum temperature of 100°C under atmospheric pressure. Use of Hydrogen peroxide for concentrated organic waste mineralisation leads to large volumes of secondary aqueous radioactive waste.

The aqueous streams with small concentrations of dissolved organics are being treated using photo oxidations methods, while Wet Air Oxidations is attractive option for concentrated organic wastes [4]. In this backdrop, it is interesting to investigate some of these concepts for mineralisation of desalination membranes further.

3. Desalination membranes and research needs

3.1 Desalination membranes

Desalination membranes are started with the cellulose acetate based materials initially. But due to their low compatibility for certain conditions such as high temperature and pH, better membranes based on aromatic polyamide became popular in later stages. To meet the large flow requirements, Thin Film composites were developed. TFCs are prepared by interfacial polymerisation on surface of porous support and operate in wide range of pH. The typical section view and surface chemistry of TFCs are as shown in **Figure 3** below.

3.2 Research needs

Various process streams in front-end/back-end of the nuclear fuel cycle are being treated at present by conventional unit operations. With available spectrum of membrane technologies today, it is pertinent to deploy these technologies either as stand-alone or as integrated processes or hybridised with conventional processes for selective separation of active species, particularly in low active process streams/wastes in addition to desalination purpose. Membranes have a definite life and deteriorates thereafter and hence unable to offer sustained quantity and quality output as desired by design. Disposal of the used membranes, hitherto not attended and cared for could prove to be a major bottleneck in the propagation of mass use of

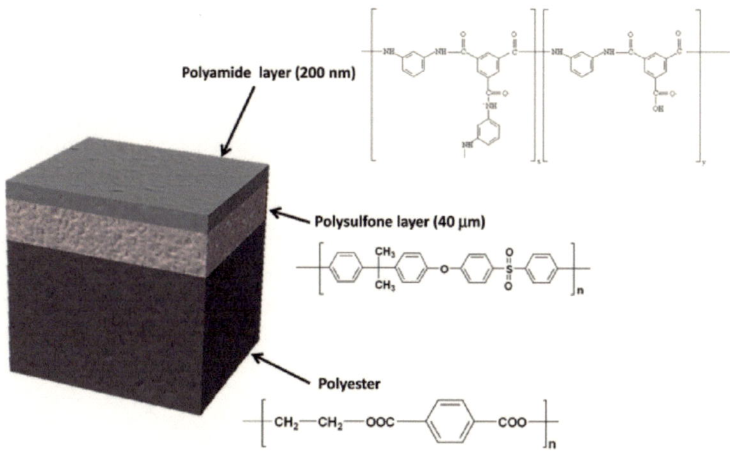

Figure 3.
Cross sectional view and surface chemistry of typical desalination membranes.

membranes as normal incineration like many other organic wastes could prove to be difficult due to complexity of off-gas treatment and generation of large volume of secondary aqueous wastes and their management. Strategies needs to be developed based on Best Available Technologies (BAT). The various polymer recycling techniques are shown in **Figure 4** below.

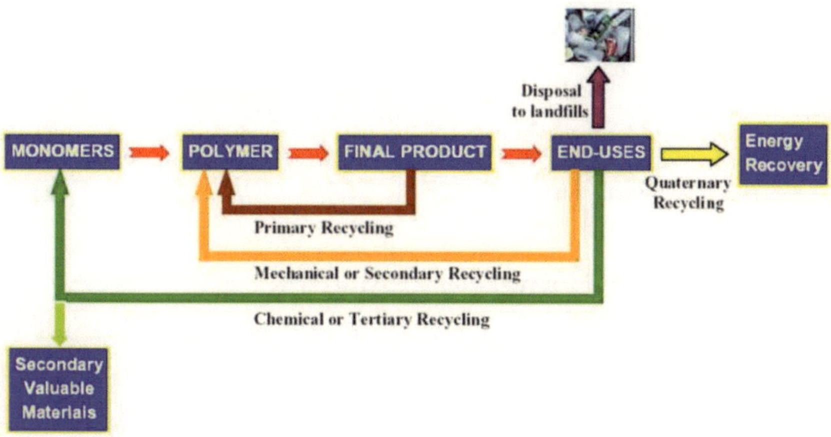

Figure 4.
Polymer recycling techniques.

4. Low temperature AOP process studies

In order to have a better insight into the process of depolymerisation of PA-TFC, the various components of the TFC membranes has been subjected to different hydro thermal process techniques and studies were performed at a constant temperature in the range of 50 to 75°C and under both static and dynamic conditions. In each run, tokens of size 20mmx20 mm with reaction solution were placed into the reactor. The set point was adjusted to a desired value, and the operation was started. At the end of run, the solution left for cooling and tokens are removed and collected for further analysis.

4.1 Effect of stirring on hydrothermal experiments (HTP-2 process technique) on unirradiated PA-TFC for different durations

Hydrothermal reaction by alkaline hydrolysis is heterogeneous two phase reaction. The intensity of turbulence in the liquid phase would decide the extent of resistance offered by the interface between the phases. Hydrothermal reaction was carried out using 50% caustic solution using polyamide TFC tokens. The reaction was carried out both under static and dynamic conditions and observed results are shown in **Figure 5**. There is an improvement in degree of depolymerisation of more than 50% under stirring speed of 330 RPM at temperature of 50°C.

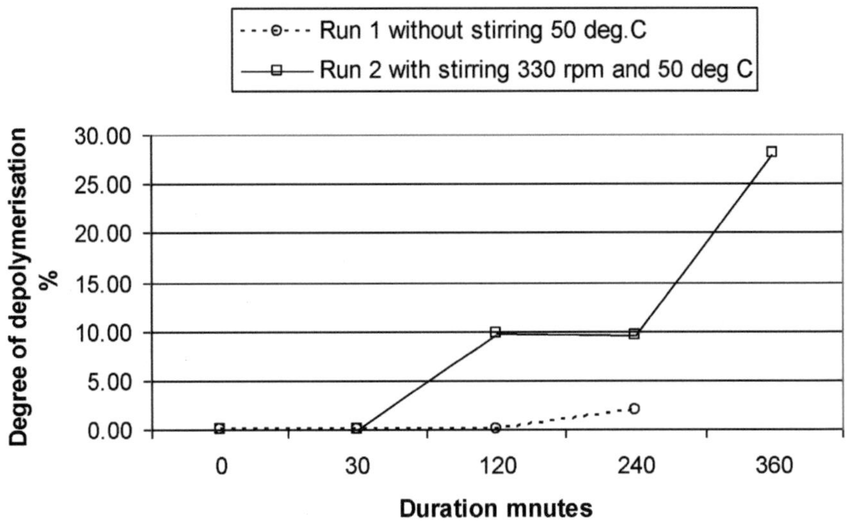

Figure 5.
Effect of stirring on HTP-2 process technique.

4.2 Effect of concentration of reaction media at 50°C

Hydrothermal reaction was carried out using 4–50% caustic solution for polyester tokens. The reaction was carried out at temperature of 50°C, under dynamic conditions, using un-irradiated polyester tokens. The observations at 50°C temperature are shown in **Figure 6**. There is an improvement in degree of depolyemrisation of more than 100% with 50% and 10% alkali concentrations. The DODP of up to 100% could be observed with 10% and 50% alkali concentrations. It is planned to carry out further studies to explore the possibility of recycling of spent polymers through selective chemical treatments.

4.3 Effect of concentration of reaction media at 75°C

Hydrothermal oxidation by alkaline hydrolysis is heterogeneous two phase reaction. Hydrothermal reaction was carried out using 4–50% caustic solution for polyester tokens. The reaction was carried out at temperature of 75°C under dynamic conditions using un irradiated polyester tokens. The results are shown in **Figure 7**. There is an improvement in degree of depolymerisation of more than 100% with 50% and 10% concentrations. The DODP of up to 100% could be observed with 10% and 50% concentraions in less than 4 hours.

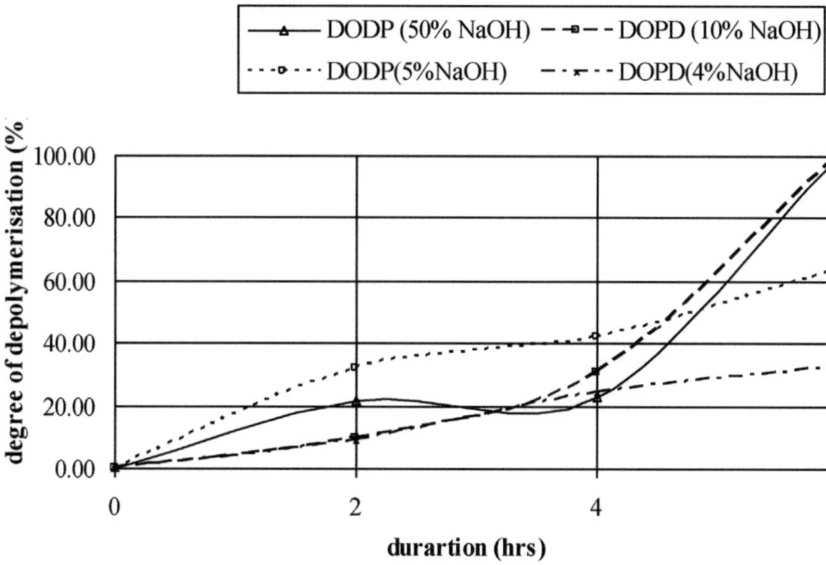

Figure 6.
Effect of concentration of reaction media on HTP-2 process technique at 50 deg. C.

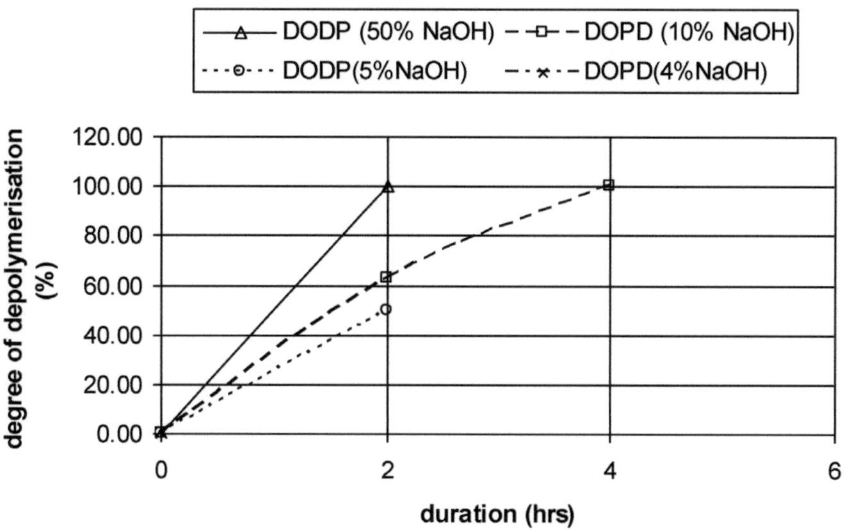

Figure 7.
Effect of concentration of reaction media on HTP-2 process technique at 75 deg. C.

4.4 Effect of irradiation on HTP-2 process technique

The purpose of this investigation is to see effect of irradiation on hydrothermal oxidation by various process techniques. Hydrothermal reaction was carried out using 4% caustic solution for polyester tokens. The reaction was carried out at temperature of 50°C under dynamic conditions, using both un-irradiated and irradiated polyester tokens. The results are shown in **Figure 8**. There is an improvement in degree of depolymerisation of more than 100% with irradiation. The DODP of up to 80% could be observed with 4% concentrations. It is planned to carry out further studies to explore the possibility of recycling of spent polymers through selective chemical and non chemical treatments.

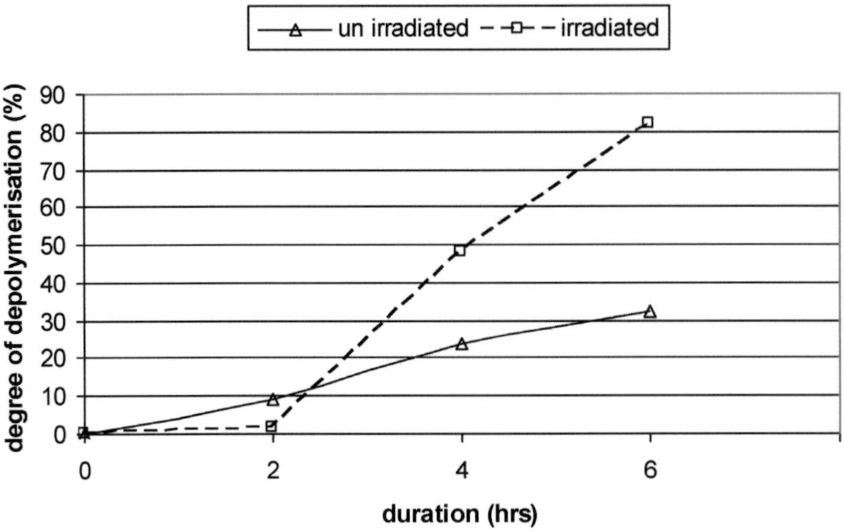

Figure 8.
Effect of radiation on HTP-2 process technique.

	Material		**Poly ester**			
	Size mm		20x20			
	Media		4% NaOH	No condensor and no refluxing		
	pH		8.90			
	Volume		50 ml			
	Temperature (°C)		50			
	Stirring rate RPM		330			
Sl. No	**Token no**	**Duration (minutes)**	**Weight loss (%)**	**COD (mg/L)**	**AchD code**	**Remarks**
1		0	0.00	0.00		
2	TK1	30	1.50	66.00	Q88	
3	TK2	60	9.94	43.00	Q89	
4	TK3	90	12.60	61.00	Q90	
5	TK4	120	16.34	51.00	Q91	
6	TK5	150	16.87	122.00	Q92	
7	TK6	180	28.55	124.00	Q93	Colour change and brittleness
8	TK7	240	36.57	113	Q94	
9	TK8	300	45.96	88	Q95	
10	TK9	360	48.8	107	Q96	

Table 2.
Variation of COD values during hydrothermal process technique w.r.t polyester.

4.5 Variation of COD values for htp-2 process techniques for desalination membrane fabric

Due to depolymerisation the total organic load will increase in the reaction media depending on the efficiency of process technique adopted. The COD/TOC gives better estimate of the same. The **Table 2** shows the variation of observed COD values in 'ppm' at an accuracy of ± 5% for hydrothermal process based on alkaline hydrolysis techniques tried on desalination membrane fabric made of polyester. The COD values of up to 120 ppm was observed and shown expected trends.

4.6 Effect of various other catalysts being studied

The non catalytic wet oxidation experiments were carried out on pure polyamide beads at temperature of 100 °C using potassium permanganate oxidation reagent for various durations. The observations at 50, 75 and 100°C are as shown below in **Figure 9**.

Further catalytic wet oxidation experiments were carried out on pure polyamide beads using 30% Hydrogen peroxide with ferrous sulphate catalyst for various durations. The observations at 50, 75 and 100°C are as shown below **Figure 10**. Further membrane performance studies are being planned to evaluate flux and salt rejection.

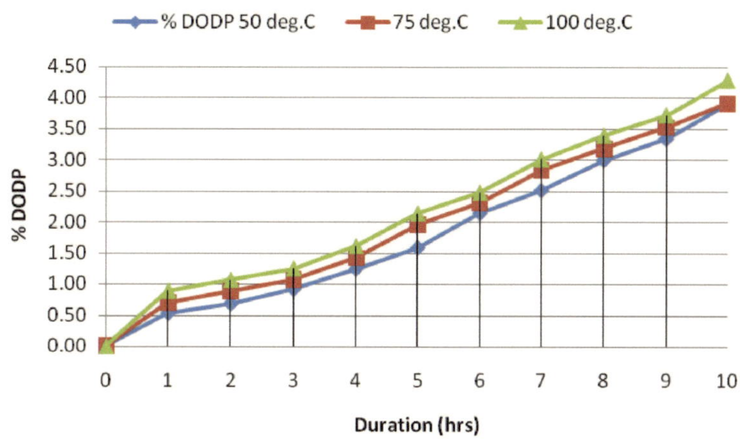

Figure 9.
Non catalytic wet oxidation observations.

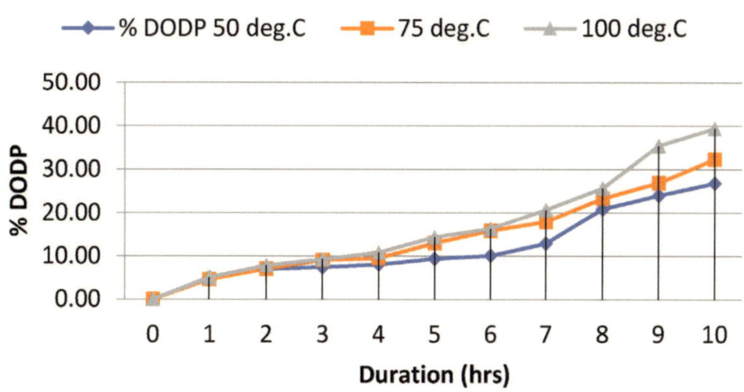

Figure 10.
Catalytic wet oxidation observations.

5. High temperature plasma studies for quaternary recycling

The recovery of plastic's energy content can be achieved through *Quaternary recycling* methods. Incineration aiming at the recovery of energy is currently the most effective way to reduce the volume of organic materials. Although polymers are actually high-yielding energy sources, this method has been widely accused as ecologically unacceptable owing to the health risk from air born toxic substances e.g. dioxins (in the case of chlorine containing polymers). To achieve improved and efficient combustion, exploratory studies were carried out using air plasma. The mass reduction factors observed for various components of desalination membranes are as shown in **Table 3** below.

The elemental composition of the residue for trail 1 with RO-TFC membranes are as shown **Table 4** below.

Operating conditions				
Voltange	160		159	volts
Current	199		200	A
Power	31.84		31.8	kW
Plasma gas flow	30		30	lpm
Additional air flow	50		200	lpm
Duration	15		20	min
Avg temperature	3000	K		
For RO-TFC membranes			For BARC water filter	
Initial mass		97		800.481 gms
Final residue mass		10.31		6.411 gms
Mass reduction factor		89.37113		99.19911

Table 3.
Air plasma studies for desalination membrane management.

Sl. No	Element	Concentration	Remarks
1	Al (%)	0.64 ± 0.05	EDXRF, ICPOES, C/S analyser, ISE, N/O analyser, H determinator
2	Cu (%)	0.61 ± 0.46	Inhomogeneous w.r.t Cu
3	Fe (%)	1.2 ± 0.2	Do
4	Si (%)	1.8 ± 0.1	Do
5	Cl^-	1.45	RSD: 5% for Cl
6	C (%)	7.0 ± 0.1	
7	S (%)	3.0 ± 0.2	
8	H (%)	0.8	RSD: 7% for H
9	N (ppm)	820	RSD: 5% for O and N
10	O (%)	44	RSD: 5% for O and N; Beyond the linear dynamic range of the technique

Table 4.
Composition of residue sample of air plasma studies.

6. Discussion

Based on Environmental regulations that have to be implemented and keeping in mind the release criteria and development of innovative futuristic radiation processing aspects, BAT (Best Available Technologies) can be foreseen. The tentative proposed logic diagram for managing solid form of wastes such as desalination membranes can be represented as shown in **Figure 11** below.

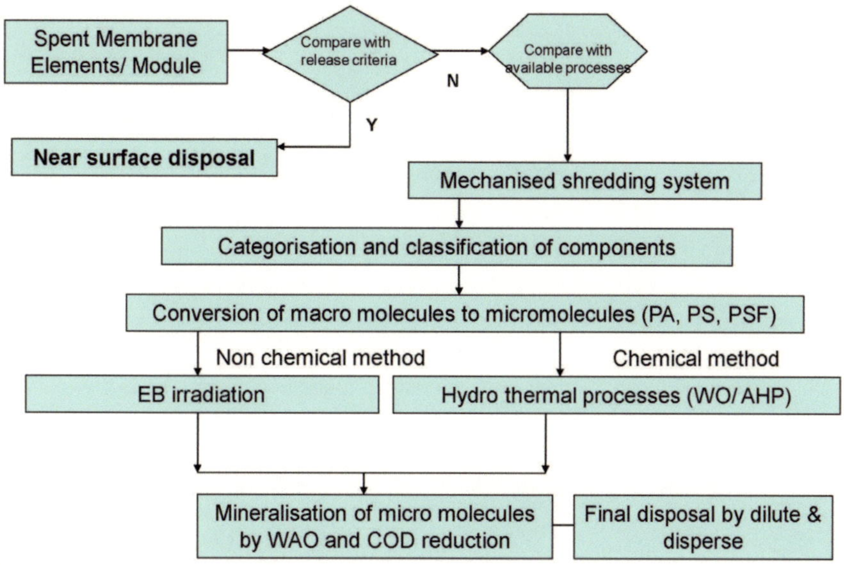

Figure 11.
Logic diagram for spent desalination membrane management.

7. Conclusions

Water industry needs to address the spent membrane management, keeping in mind to provide integrated solution. The AOPs are effective techniques to treat high degree industrial wastes. The aspects of depolymerisation and mineralisation were investigated. The variables studied have improved the degree of depolymerisation. Hybrid systems based on different techniques needs to be developed. These techniques have bright prospect in nearby future due to ongoing research initiatives. Best Available Technologies needs to be explored to supplement the conventional biological and chemical methods.

Acknowledgements

The authors wish to thank Director Chemical Engg Group for giving encouragement to the programme. Thanks are also due to various divisions of BARC for their technical suggestions and discussions. Authors wish to thank HBNI for supporting academic aspects of this programme with DGFS-PhD 2018 scholar M srija.

Conflict of interest

There is no 'conflict of interest' issues for this book chapter.

Author details

Thallam Lakshmi Prasad[1,2]

1 Desalination Division, Bhabha Atomic Research Center, Mumbai, India

2 Homi Bhabha National Institute, Bhabha Atomic Research Centre, Mumbai, India

*Address all correspondence to: tlp@barc.gov.in; tlprasad63@gmail.com

References

[1] Drioli, E., Giorno, L., Comprehensive membrane science and engineering by Elsevier publishers

[2] Treatment of spent ion exchage resin for storage and disposal, IAEA-TRS-254

[3] Prasad, T.L., Smitha Manohar., Srinivas, C., "advanced oxidation processes for treatment of spent organic resins in nuclear industry "presented at Indian chemical engineering congress, organised by Indian Institute of Chemical Engineers in collaboration with central leather research institute, Chennai, December 19-22, 2001

[4] Hwubert, D., Simon, C., and et al.," Wet Air Oxidation for treatment of industrial waste water and domestic sludges, design of bubble column reactors", Chemical Engineering Series 54(1999) 4953-4959

Chapter 3

Desalination Brine Management: Effect on Outfall Design

Ishita Shrivastava and Edward Eric Adams

Abstract

Recently proposed options for desalination brine management involve blending of brine with a lighter effluent or concentrating the brine prior to discharge, either of which can significantly alter the discharge concentrations of contaminants. We evaluate the effect of these brine management strategies on the design of submerged outfalls used to discharge brine. Optimization of outfall design is considered such that adequate mixing can be provided with minimum cost. Designs with submerged and surfacing plume are considered for outfalls located in shallow coastal regions with small currents (quiescent receiving water is assumed). Pre-dilution with treated wastewater is shown to reduce the outfall cost, whereas pre-dilution with seawater or pre-concentration are shown to result in higher costs than the discharge of brine alone. The effect of bottom slope is also explored and the results suggest that multiport diffusers are better suited than single jets at locations with a mild bottom slope.

Keywords: brine disposal, desalination, outfall, optimization, brine management, multiport diffuser

1. Introduction

Reject brine from desalination plants can have twice as high salinity as seawater [1] as well as high concentrations of other contaminants such as anti-fouling agents, anti-scalants, products of corrosion, etc., which can be harmful to benthic organisms. Thus, brine is usually discharged as a dense submerged jet which provides rapid mixing with ambient water. However, at locations that are characterized by shallow water depth and mild tidal currents, such as the north-western Arabian Gulf [2], diffusers with multiple jets are preferred as they can generate the required amount of mixing in smaller water depths.

Various options have been proposed for better management of reject brine from seawater reverse osmosis (SWRO) desalination plants [3, 4]. Processes such as pressure retarded osmosis (PRO) [3, 5] and reverse electrodialysis (RED) [6, 7] utilize the salinity difference between brine and treated wastewater effluent (TWE) to recover energy. On the other hand, processes such as electrodialysis (ED) [8] and ion-concentration polarization (ICP) [9] concentrate brine further to increase freshwater recovery [4] or lead to a zero discharge scenario. These options for brine management (pre-dilution with TWE or concentration) affect the discharge concentrations of contaminants present in brine, and can affect the design of outfall used to discharge brine.

IntechOpen

Coastal desalination plants are often co-located with power plants which provide them with low-grade heat, used in the distillation of seawater (for multistage flash desalination plants) [10], or electricity (for reverse osmosis plants). Brine is often blended with condenser cooling water (CW) from the power plant before being discharged. TWE can also be used for pre-dilution (mixing with brine before discharge) if a treatment plant is nearby. Pre-dilution helps in reducing concentrations of salt and other contaminants present in brine as well as contaminants in the pre-diluting stream (e.g., condenser cooling water or treated wastewater effluent). It also results in increased discharge flow rate (due to blending of the two streams) and reduced discharge salinity which, in turn, reduces the density of the blended effluent. This leads to progression towards shallow or vertically mixed conditions [11].

If treated wastewater effluent from a treatment plant or condenser cooling water from a coastal power plant are not utilized for pre-dilution, they are usually discharged separately and need an outfall. Thus, in addition to the reduction in discharge concentrations of contaminants, pre-dilution also leads to a reduction in total outfall cost by eliminating the need for two separate outfalls which would cost more than one outfall for the blended stream. Thus, blending of brine with cooling water or wastewater is often recommended [12].

While concentration of brine prior to discharge using submerged outfalls (which result in dilution) is not environmentally desirable in its own right, brine can be concentrated to increase freshwater recovery or harvest salts. In order to increase freshwater recovery, brine can be desalinated in two steps involving ICP and reverse osmosis (RO) [4]. ICP is used to separate brine into two streams: 1) a lighter stream with salinity of about 35 ppt, which is then desalinated using RO; and 2) a concentrated brine stream, which is either used to harvest salts or discharged using an outfall. The concentrations of contaminants present in brine increase due to concentration. Due to the high concentrations of contaminants in concentrated brine, the near-field mixing required to dilute contaminants to desirable levels is also high.

From an environmental standpoint, one is interested in reducing concentrations of contaminants in receiving water beyond a certain mixing zone. Environmental regulations usually specify the size of a mixing zone and require outfall designs that ensure that contaminant concentrations at the edge of the mixing zone are lower than specified threshold concentrations. To dilute a contaminant to a desired concentration, the outfall needs a certain water depth. At a location with offshore sloping bottom, this means going offshore to a certain distance which has an associated capital cost. Also, the cost for pumping the effluent constitutes an operating cost. The design parameters can be optimized to achieve the right balance of these two costs and design an outfall which provides desired dilutions at the end of the mixing zone with minimum cost.

We look at the effects of four brine management strategies – pre-dilution with seawater, power plant cooling water, treated wastewater effluent and pre-concentration on the design of submerged single and multiport outfalls. Outfall design variables (discharge velocity, number of ports, receiving water depth, etc.) are optimized for four different designs such that contaminants can be diluted to satisfy environmental objectives. Effect of brine management strategies on outfall cost is investigated and discussed using examples. Recommendations regarding the cost-effectiveness of different brine management options are presented.

2. Review of near-field mixing concepts for dense discharges

High velocity submerged jets are often used for the discharge of brine from desalination plants as they induce rapid mixing with ambient water and lead to reduction of contaminant concentrations. Inclined jets located near the sea floor are commonly used to discharge dense effluents as they increase the jet trajectory (and, in turn, dilution). Such jets rise to a maximum (terminal rise) height equal to y_T before the negative buoyancy causes the jets to return to the seafloor at the impact point. For a jet (with diameter D_0) discharging an effluent of density ρ_0 with a velocity of u_0 in an ambient of density ρ_a and uniform depth H, one of three regimes – deep, shallow or vertically mixed can be identified depending on the value of the shallowness parameter D_0F_0/H [11, 13]. Here, $F_0 = u_0/\sqrt{g_0'D_0}$ is the densimetric Froude number of the jet, $g_0' = (\Delta\rho/\rho_a)g = \{(\rho_0 - \rho_a)/\rho_a\}g$ is the reduced gravity and g is the acceleration due to gravity.

The receiving water is considered "deep" if its depth is sufficiently large and the dense effluent does not interact with the surface. "Shallow" conditions occur if the effluent interacts with the surface but it forms a bottom layer in the vicinity of the discharge. If the depth is small enough, the effluent can be mixed over the entire water column for large distances. Such a situation is categorized as being "vertically mixed". Increase in the value of D_0F_0/H leads to a progression towards vertically mixed conditions. For a jet inclined at 30°, the transition between deep and shallow conditions is observed at $D_0F_0/H = 0.72$ and that between shallow and vertically mixed conditions is observed at $D_0F_0/H = 7.36$ [11].

2.1 Negatively buoyant submerged jet

In deep water, the impact point dilution, which is the minimum dilution along the seafloor, of an inclined submerged jet is proportional to F_0 [14–16]. In shallow water and vertically mixed conditions, the dilution is independent of F_0 and is proportional to H/D_0 [11, 17]. The constants of proportionality depend on the discharge angle (θ_0). In deep receiving water, an inclination of 60° provides the highest dilution (for fixed value of F_0). However, smaller angles are preferred in shallow conditions [13, 17]. An inclination of 30° is chosen for further analysis which is suitable for shallow regions. For this choice of θ_0, the impact point dilutions in deep and shallow (and vertically mixed) conditions are given by Eqs. (1) and (2), respectively [11, 13].

$$S_{i,deep} = 1.2F_0 \tag{1}$$

$$S_{i,shallow} = 0.86H/D_0 \tag{2}$$

2.2 Unidirectional diffuser

A unidirectional (or tee) diffuser is an outfall which consists of an array of submerged jets (number of jets $= N$) arranged in parallel with all jets pointing in one direction perpendicular to the manifold. Use of a unidirectional diffuser is suitable in locations with mild bi-directional currents [18]. Individual jets of a unidirectional diffuser interact with each other in shallow water and lead to mixing that is different from a mere superposition of individual jets [19].

In deep water ($D_0F_0/H < 0.72$) and with adequate port spacing, there is no interaction among individual jets of a unidirectional diffuser [20] and the dilution is the same as that of a single jet (given by Eq. (1) for $\theta_0 = 30°$).

In shallow water (D_0F_0/H between 0.72 and 7.36), there is more interaction among individual jets and the impact point dilution of a unidirectional diffuser with port spacing equal to water depth ($l = H$) is given by:

$$S_{i,shallow,ud} = 0.82{F_0}^{-0.15}(H/D_0)^{1.15} \tag{3}$$

In vertically mixed conditions ($D_0F_0/H > 7.36$), the dilution is independent of the discharge buoyancy (or F_0). The impact point dilution of a unidirectional diffuser with port spacing equal to water depth ($l = H$) in vertically mixed conditions is:

$$S_{i,mixed,ud} = 0.61H/D_0 \quad (4)$$

For a unidirectional diffuser discharging in quiescent shallow or vertically mixed conditions, proximity to shoreline can result in a reduction in dilution [21]. However, the reduction in dilution is less than 15% if the separation between the diffuser and the shoreline (in constant water depth) is more than 60% of the diffuser length. At a location with uniformly sloping bottom, this is roughly equivalent to an offshore distance equal to 1.2 times the diffuser length [21]. In the presence of moderate to high crossflow, Shrivastava and Adams [22] observed no significant reduction in dilution if the separation between the diffuser and the shoreline is at least 15% of the diffuser length for a diffuser discharging in uniform water depth. This corresponds to a shoreline separation of 30% or more of the diffuser length at a location with uniformly sloping bottom.

3. Previous studies

Several studies have examined outfall optimization for brine disposal. Jiang and Law [23] provided semi-analytical solutions for the combination of port diameter (D_0) and number of ports (N) required to meet design objectives (dilution greater than a specified value and rise height of plume lower than a fraction of the water depth) for non-interfering multiport diffusers. They investigated $D_0 - N$ combinations for full submergence and surface contact scenarios (analogous to deep and shallow conditions, respectively) for a given range of brine flow rate. They did not consider a cost function but asserted that the capital cost increases with the number of ports, and thus the optimum design is the one that satisfies design objectives with minimum number of ports. They assumed jets to be non-interfering, and thus did not account for the interaction between jets in shallow water depths.

Maalouf et al. [24] provided a simulation-optimization framework to optimize SWRO outfall design. They used a regression model, calibrated using results from an initial mixing model (CORMIX), to quantify the effects of various parameters on dilution. Using this regression model for dilution, they optimized the design variables to minimize the total cost. The total cost was assumed to be a linear function of outfall pipe length (X), internal port diameter (D_0) and number of ports (N). Their analysis was based on a similar analysis done by Chang et al. [25] to evaluate optimal strategies for the expansion of a wastewater treatment plant in South Taiwan. Uncertainties in ambient parameters (e.g., ambient current speed) were also considered.

The above studies only considered linear cost functions and have not been compared to cost functions in the real world.

4. Brine management strategies

Recently proposed brine management options [3, 4] include pre-dilution with a lighter effluent and pre-concentration, and can cause significant changes to contaminant concentrations and, in turn, the required dilution. Contaminants of concern for the discharge of pre-diluted brine can be categorized into three categories

[26]. First, there are contaminants similar to salt which are present in ambient water but get concentrated due to the desalination process. Thus, the discharge concentrations are higher than ambient concentrations and these contaminants need to be diluted. Examples include salts and metals. Second, there are contaminants that are introduced by the desalination process, such as anti-scalants and cleaning chemicals [27]. Third, there are contaminants that are present in the pre-dilution stream. Examples include biochemical oxygen demand (BOD), nutrients etc. present in TWE and excess temperature from CW. While some of the contaminants of concern degrade with time (e.g., ammonia), most of them are conservative and require mixing with ambient water to reduce their concentrations below harmful levels.

For the case of pre-dilution, reject brine from a typical reverse osmosis (RO) plant (having double the salinity as ambient seawater and with flow rate $= Q_b$, reduced gravity $= {g_b}'$ and excess salinity above ambient water $= \Delta s_b$) is considered to be blended with a pre-dilution stream (flow rate $= (R_B - 1)Q_b$, reduced gravity $= {g_p}'$, excess salinity $= \Delta s_p$ and excess temperature $= \Delta T_p$), making a total flow rate of $Q_0 = R_B Q_b$. The blending ratio (R_B) is, thus, the ratio of the blended effluent flow rate (Q_0) to the brine flow rate (Q_b). The blended effluent has a reduced gravity of ${g_0}' \cong \left\{{g_b}' + (R_B - 1){g_p}'\right\}/R_B$ and excess salinity of $\Delta s_0 = \left\{\Delta s_b + (R_B - 1)\Delta s_p\right\}/R_B$. In addition to the use of TWE and CW as the pre-diluting stream, pre-dilution with ambient seawater (SW) is also considered. **Table 1** gives the properties of brine, seawater, TWE and CW used in this analysis.

Pre-dilution with TWE leads to a rapid reduction in discharge salinity as the salinity deficit of TWE (with respect to ambient water) cancels out some of the salinity excess of brine. Similarly, the reduced gravity of the effluent when brine is blended with TWE decreases rapidly. On the other hand, SW and CW do not have any salinity excess or deficit (with respect to ambient water), and thus the reduction in discharge salinity (and, in turn, reduced gravity) is less than that for the case of pre-dilution with TWE. As CW is positively buoyant with respect to ambient water, the decrease in ${g_0}'$ as a function of R_B is faster for the case of blending with CW than for the case of blending with SW.

For the case of pre-concentration, it is assumed that brine (with initial flow rate $= Q_b$) is concentrated by removing fresh water (salinity = 0 or excess salinity $= -\Delta s_b$) such that a more concentrated discharge stream is produced with flow rate of $Q_0 = Q_b/R_C$ (with $R_C > 1$). Thus, the discharge salinity is equal to $\Delta s_0 = (2R_C - 1)\Delta s_b$, where R_C is the concentration ratio defined as the ratio of the brine flow rate (Q_b) to the discharge flow rate (Q_0).

Since the salinity of brine is double the salinity of seawater and the salinity of TWE is assumed to be zero, the blended effluent has the same salinity as ambient seawater when the flows (of brine and TWE) are blended in a 1:1 ratio ($R_B = 2$). The pre-dilution of excess salinity ($= \Delta s_b/\Delta s_0$) in this case is infinite. For high values of R_B, the blended effluent may become positively buoyant ($R_B > 2$ for pre-dilution with TWE and $R_B > 8.7$ for pre-dilution with CW) in which case there is no

	Reject Brine	TWE	CW	SW
Salinity	72 ppt	0	36 ppt	36 ppt
Temperature	27°C	27°C	37°C	27°C
Reduced gravity	$0.27\ m/s^2$	$-0.26\ m/s^2$	$-0.035\ m/s^2$	0

Table 1.
Properties of brine and various pre-dilution streams.

impact point. But the dilution equations for negatively buoyant effluent are used for this case too. These results are only meant to provide qualitative predictions.

5. Optimization parameters

Optimization of the design of outfalls discharging pre-diluted or pre-concentrated brine is considered here such that regulatory requirements on contaminant concentrations can be met at the end of the mixing zone with minimum cost. The end of the mixing zone is assumed to be at the impact point of the jets. Thus, the expressions for impact point dilution of a single port outfall and a multiport (unidirectional) diffuser can be used to calculate the "physical" dilution induced by the outfall.

The location of an outfall depends on many factors, such as the availability of deep water, absence of natural submerged sills, spits, and manmade jetties, and knowledge of the offshore bathymetry; hydrodynamic modeling is often utilized to test a proposed design before it is adopted. In addition, detailed analysis of the forces exerted on the outfall due to oceanographic conditions is also carried out to ensure its stability. These factors are site-specific and beyond the scope of this chapter. Here, we are considering generic outfall designs and calculating values of design variables, such as receiving water depth, discharge velocity, number of ports, etc., that result in minimum cost. For this calculation, the outfall is considered to be located at a place with uniformly sloping bottom in the offshore direction.

Optimization of outfall design requires identification of outfall cost, desired dilution and design alternatives, which are discussed below.

5.1 Costs

One of the major components of outfall cost is the cost of the conveyance system to carry brine to the offshore discharge location. Depending on the oceanographic conditions and the discharge location, this can be done by running a pipe through a tunnel or a trench, or laying a pipe on the seabed secured using ballast weights [28]. Here, we have assumed that high density polyethylene (HDPE) pipes are used.

The capital cost is considered to be composed of four major components. The first is the cost of laying the HDPE pipe to the required offshore distance. The cost per unit length of HDPE pipes was found to be proportional to the pipe diameter (D_p) [29, 30]. Thus, the cost of the pipe is proportional to the pipe diameter (D_p) times the length of the pipe (X).

The most common way to secure HDPE pipes to the sea bed is to attach concrete ballast weights [28]. The cost of concrete weights per unit length of the pipe was found to increase with pipe diameter [29] and a linear fit was used. Thus, the total cost of anchor blocks was proportional to the product of pipe diameter and length. Combining the cost of the HDPE pipe and the concrete anchor blocks, the cost of laying the outfall pipe is:

$$CC_1 = AD_pX \tag{5}$$

At a location with uniformly sloping bottom (with slope $= \Gamma$), the length of the pipe is related to the ambient depth required ($X = H/\Gamma$). The pipe diameter depends on many factors including the size of the plant, construction material, water depth, available hydraulic head etc. [28]. Assuming the size of the pipe to be a function of the flow rate only, an analysis of the available data for outfalls around the world (from [31, 32], shown in **Figure 1**) shows the following dependence of pipe size (in m) on flow rate (in m^3/s):

$$D_p = 0.98{Q_0}^{0.36} \tag{6}$$

The cost of the outfall pipe is then given by:

$$CC_1 = a{Q_0}^{0.36}H/\Gamma \tag{7}$$

where $a = 0.98A$.

The second component is the cost of the diffuser manifold. Assuming that the diffuser manifold has the same diameter as the outfall pipe ($D_m = D_p$) and that the spacing between adjacent nozzles is equal to the water depth ($l = H$), the capital cost of the manifold becomes:

$$CC_2 = a{Q_0}^{0.36}NH \tag{8}$$

This component of cost is only considered for a multiport diffuser, i.e., $CC_2 = a{Q_0}^{0.36}H$ for a single port discharge is neglected in comparison to other costs.

The third component is the cost of nozzles. A linear fit to the cost per nozzle data, reported in [29, 30], was used to estimate the total cost of nozzles as:

$$CC_3 = N(B + CD_0) \tag{9}$$

The fourth component is the cost of pumps required to pump the effluent to the offshore location of the outfall. The cost of pumps increases with the flow rate and the total head loss in the outfall. Based on the cost of pumps for pumping product water reported by [29], this cost was found to be proportional to the product of effluent density, flow rate and total head loss (H_L). Thus:

$$CC_4 = E\rho_0 Q_0 H_L \tag{10}$$

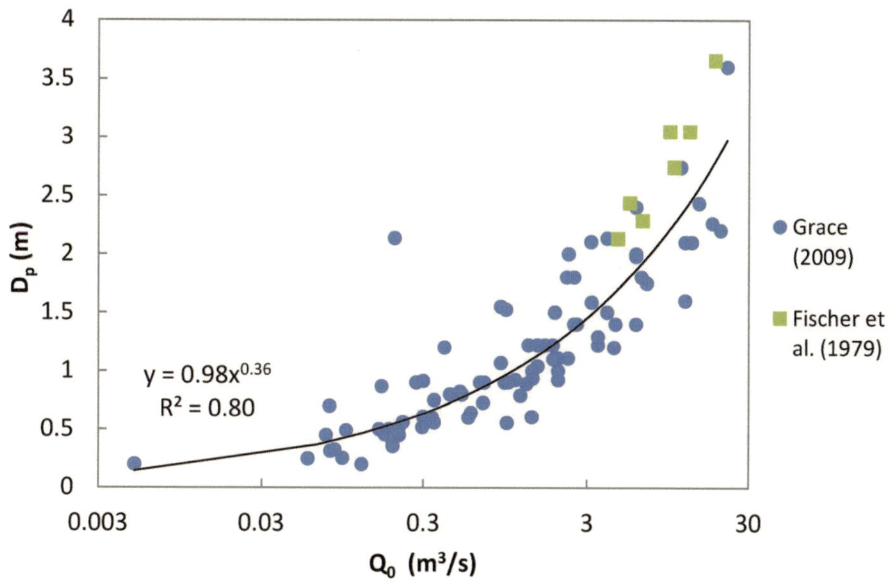

Figure 1.
Correlation between outfall pipe diameter and flow rate.

The first three cost components (CC_1, CC_2 and CC_3) only include material costs. The installation cost is assumed to be 1.2 times the material cost (based on cost estimates from [30]) so that the total cost is 2.2 times the material costs. For the cost of pumps (CC_4), the installation cost is already included in Eq. (10).

The total cost of the outfall also includes an operating cost which mainly consists of the cost of electricity for pumping the effluent, and operation and maintenance cost. It is assumed that the available pressure and elevation head before discharge are negligible and thus pumping is required to discharge the effluent with high velocity. The pumping cost is proportional to the product of effluent density, flow rate and total head loss. Thus, the pumping cost over the life of the plant is:

$$OC_1 = F\rho_0 Q_0 H_L \tag{11}$$

where F depends on the cost of electricity, discount rate and outfall lifetime.

Malcolm Pirnie [29] reported values of operation and maintenance cost for different scenarios which suggest that it is independent of design variables. Therefore, a constant value was used for the operation and maintenance cost.

Table 2 provides a summary of the cost functions and typical values of cost coefficients (for costs in USD, as per May 2016 ENR index).

An estimation of head loss is required to calculate the total cost. Head loss is estimated by considering the components listed in **Table 3**. Here, V_p is the velocity inside the outfall pipe. The head loss incurred in conveying the effluent to the shoreline is not included as it is the same for all designs and does not affect the optimization analysis. Thus, the outfall costs calculated here represent the cost above the cost of the simplest (shoreline) discharge.

5.2 Desired dilution

Environmental regulations usually specify threshold concentrations for various contaminants. These are maximum acceptable concentrations in the water body that are considered to be safe for aquatic organisms. Thus, outfalls are required to reduce contaminant concentrations to threshold levels within a regulatory mixing zone. Here, the impact point of the jets is assumed to be the end of the mixing zone.

Threshold concentrations can be different at different locations as they are based on the toxicological adaptability of the marine species thriving in that location. Also, regulatory requirements vary from country to country, with international guidelines also referring to local regulations [34, 35]. In addition, source stream concentrations vary depending on the quality of feed water, desalination process etc.,

Costs	Expression	Cost coefficients
Cost of outfall pipe	$CC_1 = 2.2aQ_0{}^{0.36}H/\Gamma$	$a = 1.47 \times 10^3$
Cost of diffuser manifold	$CC_2 = 2.2AD_mNH$	$A = 1.5 \times 10^3$
Cost of nozzles	$CC_3 = 2.2N(B + CD_0)$	$B = 2.1 \times 10^3, C = 3.3 \times 10^4$
Cost of pumps	$CC_4 = E\rho_0 Q_0 H_L$	$E = 35$
Pumping cost[a,b]	$OC_1 = F\rho_0 Q_0 H_L$	$F = 73$
Operation and maintenance cost[b]	$OC_2 = G$	$G = 1.4 \times 10^6$

[a]*Assuming cost of electricity = 0.10 USD/kWh.*
[b]*Assuming discount rate of 10% and plant lifetime of 20 years.*

Table 2.
Break-down of total outfall cost.

Component	Description	Expression	Coefficient value
Conveyance to offshore location of the outfall	Friction loss in a pipe of length X and diameter D_p	$f(X/D_p)(V_p{}^2/2g)$	$f = 0.015$
A T-junction[a]		$K_T(V_p{}^2/2g)$	$K_T = 1$[d]
Diffuser manifold	Friction loss in a pipe of length $L = NH$ and diameter D_p	$f(NH/D_p)\{(V_p/2)^2/2g\}$	
Entry loss[a]	Loss incurred while entry into the riser[c]	$K_{en}(u_0{}^2/2g)$	$K_{en} \approx 0.3$[e]
Sudden contraction[b]	Contraction from pipe diameter to nozzle diameter	$K_c(u_0{}^2/2g)$	K_c[f]
A 30° elbow	For the nozzles pointing at 30°	$K_{el}(u_0{}^2/2g)$	$K_{el} = 0.3$[d]
Exit loss		$u_0{}^2/2g$	

[a] *Only for the design with a unidirectional diffuser.*
[b] *Only for a single port design.*
[c] *Assuming riser diameter equal to the nozzle diameter.*
[d] *From Davis [33].*
[e] *$K_{en} \approx 0.2 + (V_d/V_r)^2$ from Fischer et al. [31]. V_d is the velocity inside the manifold and V_r is the velocity in the riser. With the constraint on u_0 for uniform flow (discussed later), K_{en} has a maximum value of 0.3 which is used here.*
[f] *Assumed to vary linearly with the ratio of cross-sectional areas of the two pipes from 0.45 to 0.16 for area ratio (ratio of smaller cross-sectional area to larger cross-sectional area) from 0.04 to 0.64.*

Table 3.
Components of total head loss.

resulting in a range of values of the desired dilution. For simplicity, salinity is assumed to be the most constraining contaminant. The threshold concentration of salt is assumed to be 2 ppt in excess of ambient salinity [36] and outfall designs which dilute salinity to an excess of 2 ppt at impact point are discussed.

Effective dilution for a contaminant is defined as the ratio of its excess concentration in the source stream (e.g., brine for salinity) to its excess concentration at a given location. Thus, if the excess salinity of the diluted effluent at the impact point is equal to 4 ppt (in excess of ambient salinity), then the effective dilution of salinity at impact point is equal to $36/4 = 9$, where 36 ppt is the excess salinity of reject brine (**Table 1**). Similarly, the desired effective dilution for any contaminant is the ratio of its concentration in the source stream to the threshold concentration (both in excess of ambient concentration). Thus, the desired effective dilution of salinity is equal to 18.

Unlike the desired effective dilution, the desired physical dilution at the impact point also depends on the amount of pre-dilution or pre-concentration (the value of R_B or R_C), in addition to the source streams and threshold concentrations. For example, if brine is pre-diluted with TWE with $R_B = 1.5$, then the discharge excess salinity is 12 ppt and the desired physical dilution is equal to $12/2 = 6$, which is different than the desired effective dilution which is equal to 18. The outfall design in this case needs to provide an impact point dilution of 6.

5.3 Design alternatives

Brine can be discharged through an outfall in two ways – the discharge can be such that the plume stays below the water surface or it can be allowed to hit the surface. The former design would be implemented if the regulations require the plume to not be visible at the surface. However, the latter design usually costs less and should be preferred when there are no restrictions on plume visibility.

For a jet inclined at 30°, the depth below which the impact point dilution is affected by the water surface is more than the depth at which the jet hits the water surface [11]. Thus, for a submerged plume (which is not allowed to hit the surface), the maximum dilution (with minimum total cost) is achieved when the terminal rise height of the jet is just high enough that the ambient depth affects the dilution, i.e., at the transition point between deep and shallow conditions ($D_0 F_0/H = 0.72$). To dilute a contaminant to a threshold concentration, the design variables for this design can be determined by ensuring that the physical dilution is just enough to get the desired concentration and the discharge plume rises to just below the water surface ($D_0 F_0/H = 0.72$ for an inclination of 30°). The design variables for this design are denoted using the subscript 'd', for deep.

These design parameters do not minimize the total cost as they require a large capital cost. Specifically, in locations with very small bottom slope, such as the Arabian Gulf [2], the capital cost can be several orders of magnitude larger than the pumping cost and the total cost can be very high. To reduce the capital cost, it is beneficial to achieve the desired dilution with smaller ambient depth by reducing the port diameter or to employ a multi-port diffuser. Using a single, smaller diameter port will result in an increase in discharge velocity, and thus the pumping cost.

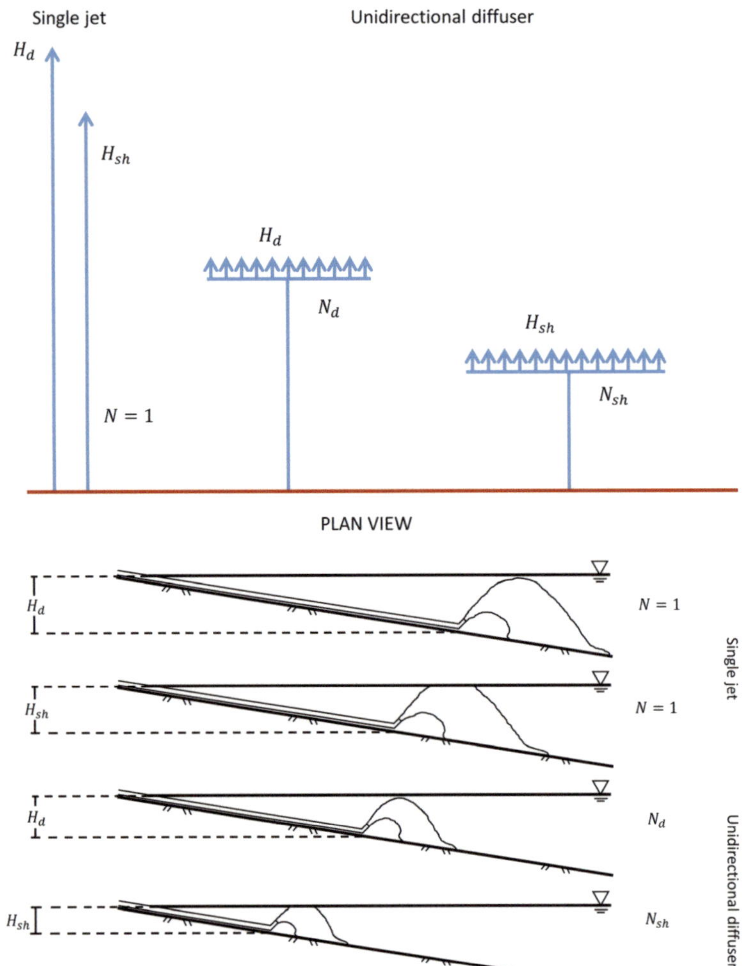

Figure 2.
Schematic showing the plan view (top) and elevation view (bottom) of the four designs considered.

The optimum design will be the one that minimizes the total cost (capital cost + pumping cost). The design variables for this design are denoted using the subscript 'sh', for shallow. Similarly, for a multiport diffuser, optimum design variables can be computed for the two designs, one with the diffuser plume submerged and the other with surfacing plume. A schematic of the four designs is shown in **Figure 2**.

6. Design optimization

6.1 Discharge through a single jet creating a submerged plume

The optimum values of water depth, diameter and discharge velocity needed to dilute a contaminant with excess concentration of Δc_O to a desired excess concentration of Δc_{th}, with the additional constraint that the plume remains submerged, are given by Eqs. (12)–(15), respectively. H_d depends on the mass loading of the contaminant (in excess of ambient concentration, $\dot{m} = Q_0 \Delta c_O$), buoyancy flux of the effluent ($B_O = Q_0 {g_0}'$) and desired concentration (Δc_{th}), but is independent of the flow rate (Q_0) as shown in Eq. (13). Therefore, the required water depth for salinity as the contaminant of concern and seawater as the pre-diluting stream is independent of R_B (as $\dot{m}$ and B_O are independent of R_B in that case).

$$H_d = 1.38 \frac{{Q_0}^{2/5}}{{g_0}'^{1/5}} \left(\frac{\Delta c_O}{\Delta c_{th}} \right)^{3/5} \tag{12}$$

$$\text{or } H_d = \frac{1.38}{{B_O}^{1/5}} \left(\frac{\dot{m}}{\Delta c_{th}} \right)^{3/5} \tag{13}$$

$$(D_O)_d = 1.18 \frac{{Q_0}^{2/5}}{{g_0}'^{1/5}} \left(\frac{\Delta c_{th}}{\Delta c_O} \right)^{2/5} \tag{14}$$

$$(u_O)_d = 0.91 {Q_0}^{1/5} {g_0}'^{2/5} \left(\frac{\Delta c_O}{\Delta c_{th}} \right)^{4/5} \tag{15}$$

Figure 3 shows the variation of H_d, $(D_O)_d$ and $(u_O)_d$ as functions of R_B (for different pre-dilution streams) and R_C. The variables are scaled so that they can be plotted on the same plot. The scaling is the same for all the pre-dilution cases (indicated in the legend for SW blending plot) but is different for the pre-concentration case (indicated in the legend for pre-concentration plot) because of the different range of values.

When brine is pre-diluted, the desired physical dilution reduces with an increase in R_B (except for $R_B > 2$ for blending with TWE), and thus the discharge velocity also reduces. For the case of brine concentration, the desired physical dilution increases rapidly as R_C increases and the effluent needs to be discharged with very high velocity to achieve better mixing. For example, the desired physical dilution is equal to 54 for $R_C = 2$ and $(u_O)_d$ is equal to 17.7 m/s which is not realistic. $(u_O)_d$ is even higher for higher values of R_C which suggests that a single jet should not be used to discharge concentrated brine at a location with restriction on plume visibility.

6.2 Discharge through a single jet creating a surfacing plume

This section explores the optimum design with no restriction on plume visibility, i.e., the design which minimizes total cost without any constraint. For most cases, this design results in a plume which hits the surface. But for some cases, the design

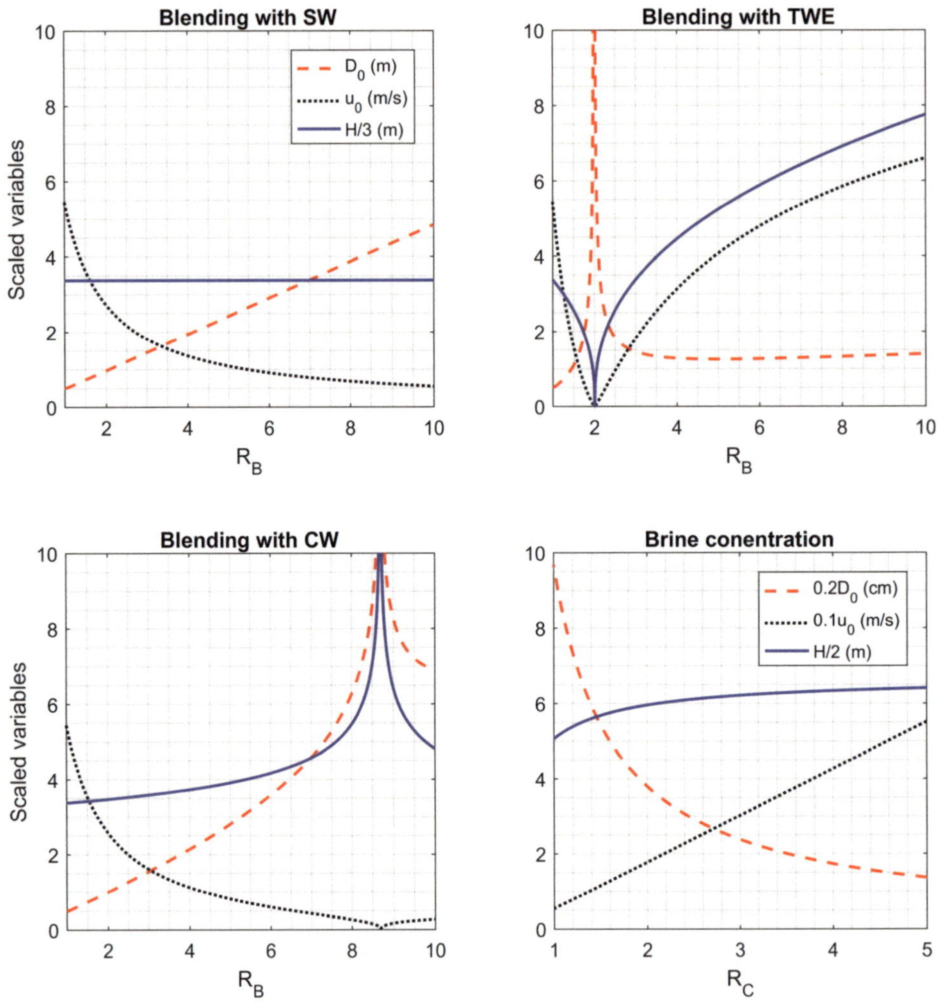

Figure 3.
Variation of H, D_0 and u_0 with R_B and R_C for discharge using a single jet with submerged plume for $Q_b = 1\ m^3/s$, $\Gamma = 0.01$ and a desired excess salinity of 2 ppt. (variables are scaled differently for the pre-dilution and pre-concentration cases as indicated in the legend).

with a submerged plume is also the one which minimizes the total cost and should be adopted. This design optimization results in non-linear equations which are solved using the 'fsolve' function in MATLAB.

Figure 4 shows the variation of H_{sh}, $(D_0)_{sh}$ and $(u_0)_{sh}$ as functions of R_B and R_C. Unlike the design with a submerged plume where the required water depth is either constant or increases with R_B (for pre-dilution with SW and CW), the required water depth for the surfacing plume design reduces with R_B as the desired physical dilution reduces. For pre-dilution with TWE, the required water depth follows the same trend as the desired physical dilution. Thus, it reduces with R_B for $R_B < 2$ and increases with R_B for $R_B > 2$. When brine is concentrated, the design with a submerged plume is the optimum design for $R_C > 1.4$ because the smaller flow rate and higher density difference (as compared to brine which is not concentrated) are less likely to lead to shallow conditions. Thus, even when there are no restrictions on plume visibility, the design of a single jet to discharge concentrated brine results in unrealistically high values of u_0. For all cases (except pre-concentration with $R_C > 1.4$), the design with a surfacing plume has a higher discharge velocity and lower water depth than the corresponding values for the design with submerged

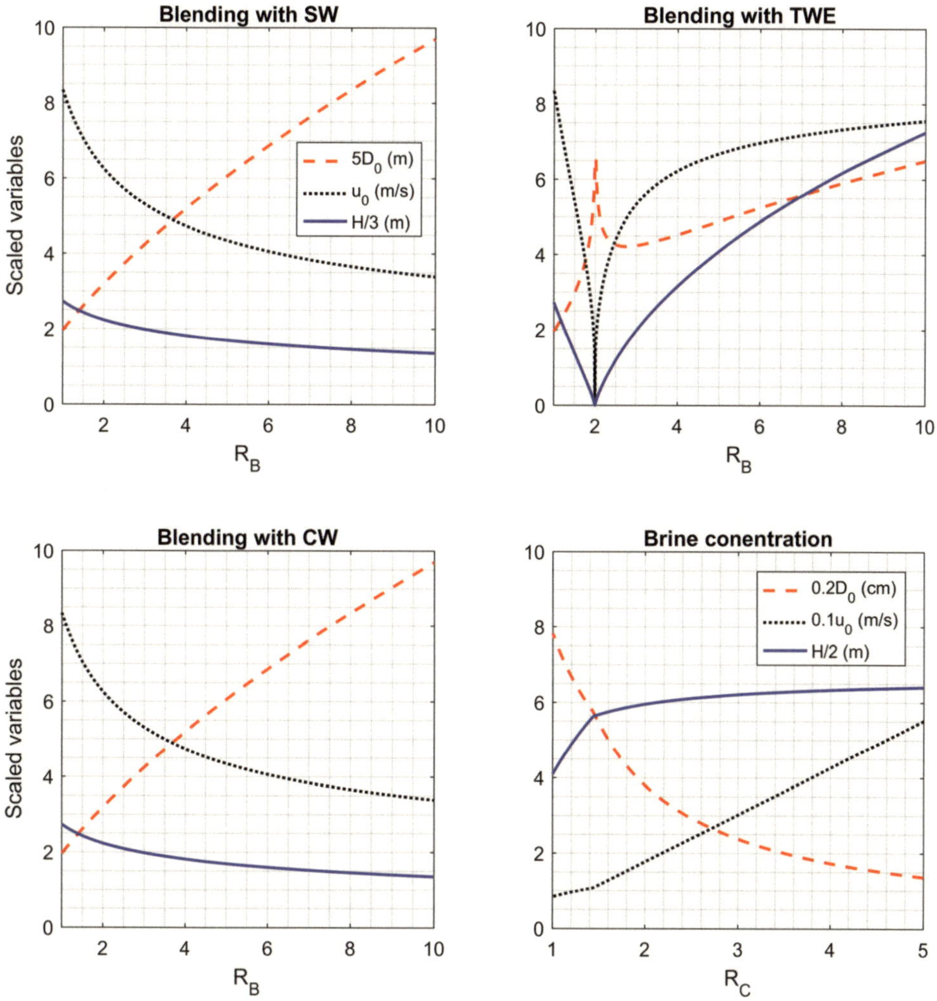

Figure 4.
Variation of H, D_o and u_o with R_B and R_C for discharge using a single jet with surfacing plume for $Q_b = 1\ m^3/s$, $\Gamma = 0.01$ and a desired excess salinity of 2 ppt. (variables are scaled differently for the pre-dilution and pre-concentration cases as indicated in the legend).

plume. The higher velocity helps in generating the same amount of mixing as the submerged plume case but in smaller water depth.

6.3 Discharge through a unidirectional diffuser

The design optimization for a unidirectional diffuser also results in non-linear equations which are solved using the 'fsolve' function in MATLAB. Optimum design variables are calculated which achieve desired dilution and minimize total cost. However, in some cases the optimized design variables need to be adjusted. For example, to ensure uniform flow through all the ports, the aggregate cross-sectional area of the nozzles should be less than two-thirds of the cross-sectional area of the diffuser manifold [31]. Since the manifold diameter is assumed to be related to the discharge flow rate (Eq. (6)), this requires the discharge velocity to be at least equal to $2Q_0^{0.28}$. Thus, if the optimum value of u_0 is less than $2Q_0^{0.28}$, u_0 is fixed to be equal to $2Q_0^{0.28}$ and other design variables are re-evaluated to minimize total cost.

For certain cases, the design with a single port is the one which minimizes cost, i.e., any design with multiple ports will have higher total cost than the design with one port. This is observed for cases which require a submerged plume and for which the desired physical dilution is small. The optimum discharge velocity (not adjusted for uniform flow) for such cases is small and adjustment for uniform flow results in a design with total cost higher than the cost of the single jet design. For these cases, the single port design is accepted as the optimum design.

Once the optimum design variables are calculated (which satisfy all constraints), N is rounded to the nearest integer and D_0 is adjusted such that $Q_0 = (\pi/4)Nu_0{D_0}^2$.

6.3.1 Discharge through a unidirectional diffuser creating a submerged plume

Figure 5 shows the variation of H_d, $(D_0)_d$, $(u_0)_d$ and N_d as functions of R_B and R_C. The design with a single jet is the optimum design for $R_B > 2.1$ for blending with SW and CW, and for R_B between 1.4 and 3.8 for blending with TWE. The discharge velocity is fixed to be equal to $2{Q_0}^{0.28}$ (to ensure uniform flow through nozzles) for

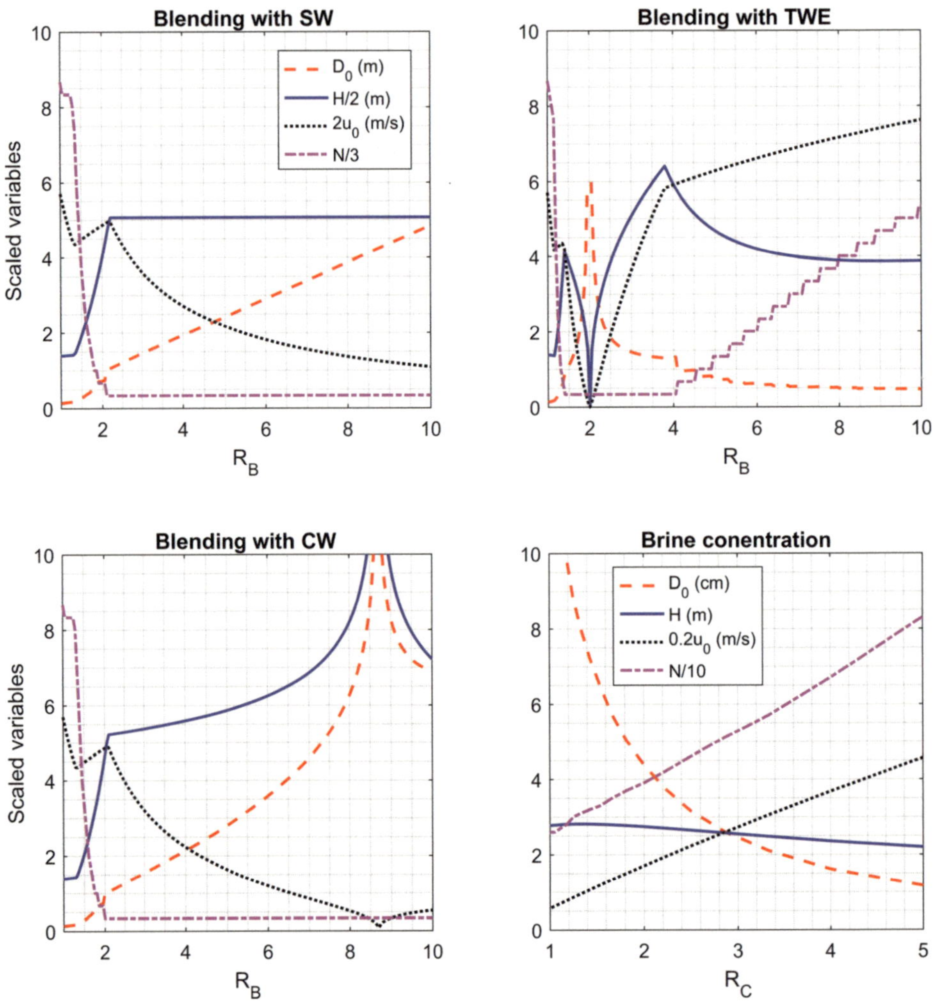

Figure 5.
Variation of H, N, D_o and u_o with R_B and R_C for discharge using a unidirectional diffuser with submerged plume for $Q_b = 1\ m^3/s$, $\Gamma = 0.01$ and a desired excess salinity of 2 ppt. (variables are scaled differently for the pre-dilution and pre-concentration cases as indicated in the legend).

R_B between 1.3 and 2.1 (for pre-dilution with SW and CW), and for R_B between 1.2 and 1.4 and greater than 3.8 (for pre-dilution with TWE).

6.3.2 Discharge through a unidirectional diffuser creating a surfacing plume

An optimum design with multiple ports (which has lower cost than a single port design) can be found for all cases when the effluent plume is allowed to hit the surface. **Figure 6** shows the variation of H_{sh}, $(D_0)_{sh}$, $(u_0)_{sh}$ and N_{sh} as functions of R_B and R_C. The discharge velocity needs to be adjusted to ensure uniform flow for $R_B > 4.2$ when brine is blended with SW and CW and for $R_B > 9.6$ when brine is blended with TWE. For other cases, all variables can be adjusted to minimize cost. The required water depth can be seen to reduce as R_B increases for pre-dilution with SW and CW. This is similar to the case of a single jet with surfacing plume.

For the multiport diffuser designs calculated here, the ratio of offshore distance of the diffuser (X) to its length (L) is more than 3 for the pre-dilution cases and more than 1.2 for the pre-concentration cases. For these values of X/L, the presence

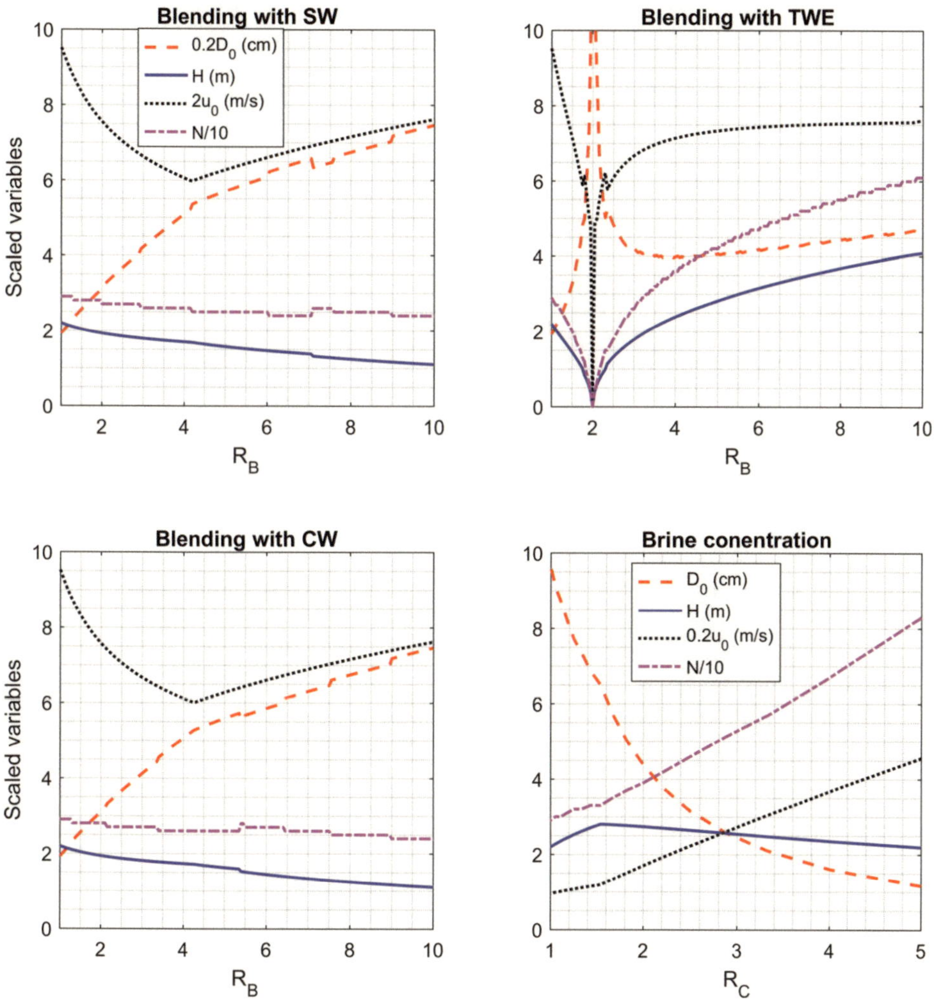

Figure 6.
Variation of H, N, D_0 and u_0 with R_B and R_C for discharge using a unidirectional diffuser with surfacing plume for $Q_b = 1\ m^3/s$, $\Gamma = 0.01$ and a desired excess salinity of 2 ppt. (variables are scaled differently for the pre-dilution and pre-concentration cases as indicated in the legend).

of the shoreline is not expected to have a significant effect on outfall dilution (dilution reduction of less than 15% in stagnant receiving water) [21, 22].

7. Results and discussion

7.1 Cost of outfalls

Figure 7 shows the comparison of total costs for the four designs (single jet and unidirectional diffuser with submerged and surfacing plume) with $Q_b = 1\ m^3/s$, $\Gamma = 0.01$ and desired excess salinity of 2 ppt. It can be seen that for pre-dilution with SW and CW, the costs of all four designs increase with increase in R_B in spite of the fact that the desired physical dilution decreases with increase in R_B. The increase in total cost is caused due to the increase in discharge flow rate. For the case of blending with TWE, however, the costs of all four designs decrease with increasing R_B for $R_B < 2$ due to the rapid reduction of desired physical dilution in that case. (The desired physical dilution goes down from 18 for $R_B = 1$ to 6 for $R_B = 1.5$.) The blended effluent is positively buoyant for $R_B > 2$ and $R_B > 8.7$ when brine is blended with TWE and CW, respectively, and the trends shown are

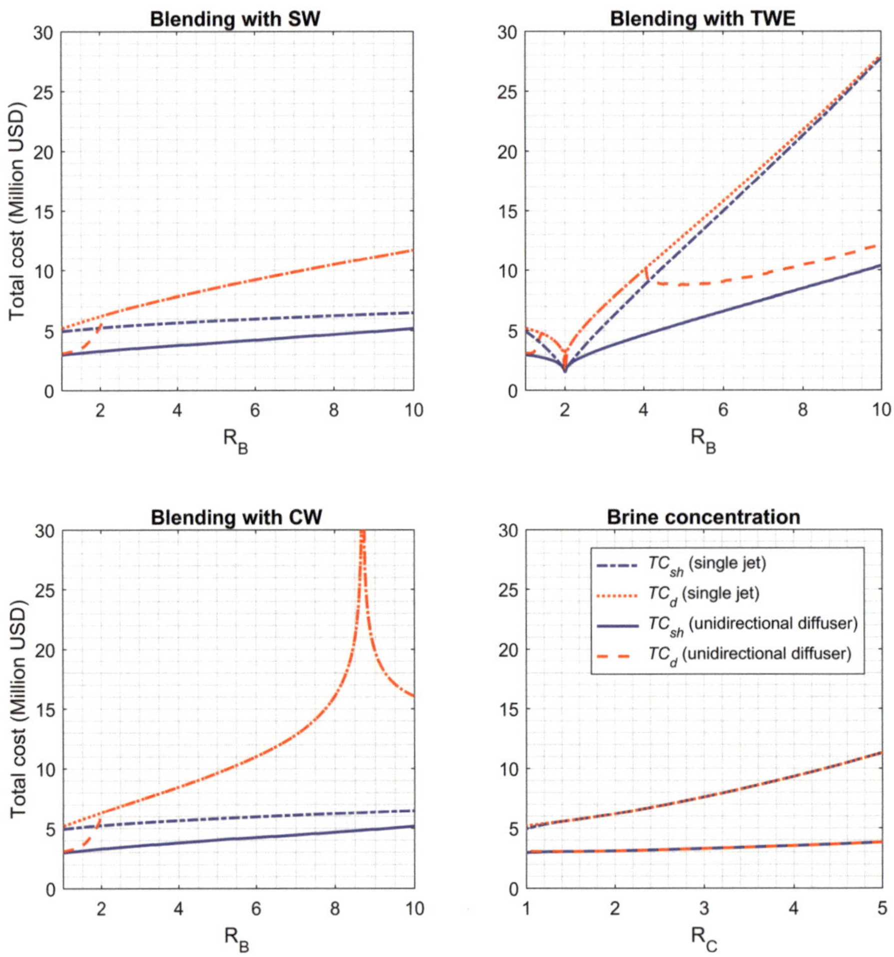

Figure 7.
Total costs of the four design alternatives to achieve desired excess salinity of 2 ppt at the impact point with $Q_b = 1\ m^3/s$ *and* $\Gamma = 0.01$.

different for R_B in this range. For the case of pre-concentration, the desired physical dilution increases rapidly with R_C leading to the increase in total cost.

Figure 7 shows that for most of the pre-dilution cases, the design with a single jet is the optimum design when the regulations require the plume to be submerged. Thus, for these cases, the 'TC_d (single jet)' and 'TC_d (unidirectional diffuser)' curves overlap. For pre-dilution cases, the total cost can be significantly lower for the surfacing plume design as compared to the submerged plume design. For blending with SW ($R_B = 2$), TWE ($R_B = 1.5$) and CW ($R_B = 5$), the ratio of the total cost for surfacing plume design to that for the submerged plume design with a single jet is 0.84, 0.77 and 0.60, respectively. Using a unidirectional diffuser, this ratio is 0.60, 0.58 and 0.42 for blending with SW ($R_B = 2$), TWE ($R_B = 1.5$) and CW ($R_B = 5$), respectively.

For the discharge of brine without pre-dilution or pre-concentration, the total costs (in million USD) of the four designs are $TC_d = 5.2$ and $TC_{sh} = 4.9$ (for a single jet discharge), and $TC_d = 3.1$ and $TC_{sh} = 3.0$ (for a unidirectional diffuser). Thus, compared to the cost of a single jet design with submerged plume, the cost can be reduced by 40% if a multiport diffuser is used (with submerged plume), by 6% if the plume is allowed to hit the surface (but still using a single jet), and 42% if a multiport design with a surfacing plume is adopted.

When brine is concentrated, the desired physical dilution increases rapidly with increase in R_C. Hence, discharge of concentrated brine is not preferable from an environmental standpoint. If brine is concentrated, it needs to be discharged with high (perhaps unrealistic) discharge velocity and/or using a large number of ports to generate adequate mixing. **Table 4** shows an example of the design variables calculated for $R_C = 2$. (Only designs with submerged plume are included because they are also the designs which minimize cost).

Pre-concentration of brine increases the concentrations of contaminants present in brine. Thus, the total cost of discharging concentrated brine increases with R_C as shown in **Figure** 7. The processes used to concentrate brine also have some cost. Thus, whether brine should be concentrated prior to discharge depends on the value of the extra fresh water produced compared to the cost of pre-concentration and the additional cost of the outfall. At locations with regulatory restrictions on discharge concentrations, pre-concentration might not be possible. Pre-concentration could be beneficial if brine is concentrated to the extent that salts can be crystallized as there would be no cost of discharge.

The costs in **Figure** 7 are calculated for salinity as the contaminant of concern. However, the relative importance of different types of contaminants (present in brine, TWE or CW) depends on the blending ratio (for pre-dilution with TWE and CW). At low blending ratio, the contaminants present in brine require higher dilution and are likely to be the constraining contaminants whereas contaminants present in TWE or CW require higher dilution at high blending ratio. Thus, the designs and the associated costs calculated above need to be adjusted at high blending ratio.

Design	N	H (m)	u_0 (m/s)	Capital cost (million USD)	Operating cost (million USD)	Total cost (million USD)
Single jet with submerged plume	1	11.9	17.7	3.6	2.6	6.2
Unidirectional diffuser with submerged plume	39	2.7	8.5	1.4	1.7	3.1

Table 4.
Example showing calculated design variables for the discharge of concentrated brine ($R_C = 2$) with $\Gamma = 0.01$.

7.2 Effect of threshold concentrations on outfall design

Since the outfall design depends on desired physical dilution, which in turn, depends on the threshold concentrations, it is important to analyze the effect of threshold concentrations on the optimum design. This is illustrated through an example in **Figure 8** in which the threshold concentration of salinity (Δs_{th}) varies between 0.5 and 5 ppt (above ambient). The variation in required depths and total costs with the threshold salinity is shown for discharge of brine without pre-dilution or pre-concentration.

The required depths and total costs (for designs with submerged and surfacing plume) decrease with increase in threshold concentrations (for discharge through a single jet) because the additional mixing required to achieve those concentrations is less. For a design with multiple ports which requires the plume to be submerged and has the discharge velocity fixed to ensure uniform flow, the required depth is proportional to the inverse of desired dilution, i.e., the depth is proportional to Δs_{th}. This can be seen for $\Delta s_{th} > 3.3$ ppt. When the discharge velocity is not fixed (for $\Delta s_{th} < 3.3$ ppt), the required depths and total costs reduce with increase in Δs_{th} similar to the case of a single jet discharge.

7.3 Effect of bottom slope

The optimum design at a location with a mild bottom slope, such as the Arabian Gulf which has bottom slopes as little as about 4×10^{-4} [2], can be quite different as compared to the design at a location with a steep slope. With a mild bottom slope, the offshore distance to locate the outfall in sufficient depth of water can be long which also increases the total cost significantly. In that case, it costs less to achieve the desired dilution in small water depth by increasing the discharge velocity and/or the number of ports. This is illustrated by considering outfall designs at two locations with $\Gamma = 0.01$ and 0.001. For discharge using a single jet in deep water (submerged plume), the design variables are independent of Γ (Eqs. (12)–(14)) but the total cost is higher for a location with smaller bottom slope because of the increased offshore distance. For discharge using a single jet with a surfacing plume and discharge through a unidirectional diffuser, the design variables can be adjusted to reduce the total cost. But, the total costs are still significantly higher for smaller Γ. **Figure 9** shows the effect of Γ on the total cost to discharge brine pre-diluted with SW using a single jet and a multiport diffuser. The total cost for a submerged plume

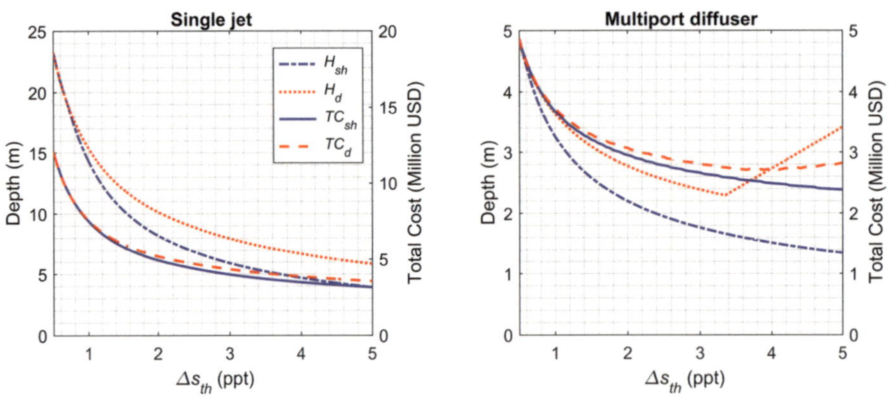

Figure 8.
Variation of H_d, H_{sh}, TC_d and TC_{sh} with threshold salinity for discharge of brine through a single jet and a tee diffuser with $Q_b = 1\ m^3/s$ and $\Gamma = 0.01$.

design with $\Gamma = 0.001$ is approximately 10 times the corresponding cost for $\Gamma = 0.01$.

A comparison of optimum design variables at locations with different bottom slopes is shown in **Table 5** for discharge of brine without pre-dilution or pre-concentration. For this example, two bottom slopes ($\Gamma = 0.01$ and $\Gamma = 0.001$) are considered. The design of a single jet with surfacing plume for $\Gamma = 0.001$ has a discharge velocity of 20.8 m/s which is not realistic. The designs with multiple ports are preferable with reasonable velocities. It can be seen from **Table 5** that the cost of the unidirectional diffuser design is about 60% of the cost of a single jet design for $\Gamma = 0.01$ but only 25% of the single jet cost for $\Gamma = 0.001$ which suggests that a multiport design is more realistic at locations with small Γ.

For the unidirectional diffuser designs in **Table 5**, the required water depths are 1.4 m and 0.8 m (for $\Gamma = 0.001$). Thus, the lengths of outfall pipe to outfall location are 1.4 km and 0.8 km, which are quite long. For such locations, a staged diffuser [37] can also be used which has ports along the length of the outfall pipe. For the same diffuser length, water depth, flow rate and discharge velocity, the dilution of a staged diffuser in quiescent conditions is less than the dilution of a unidirectional diffuser [18]. But considering that the length of the outfall pipe is much longer as compared to the diffuser length for a unidirectional diffuser design, the staged diffuser design will get much higher dilution than the unidirectional diffuser. In fact, if a staged diffuser is designed to achieve the desired physical dilution, its offshore distance would be less that the 1.4 km (or 0.8 km) distance for the unidirectional diffuser design.

7.4 Comparison with the cost of discharging brine without pre-dilution or pre-concentration

As shown in **Figure 7**, the cost of discharging brine blended with TWE is less than the cost of discharging brine without pre-dilution for $R_B < 2$. However, the total costs (for all four designs) for other pre-dilution cases increase as R_B increases (except when brine is blended with CW with $R_B > 8.7$), which means that the cost of discharging pre-diluted brine is higher than the cost of discharging brine without pre-dilution. However, these costs should be compared to the cost of two outfalls for discharging brine and the pre-dilution stream separately (for blending with

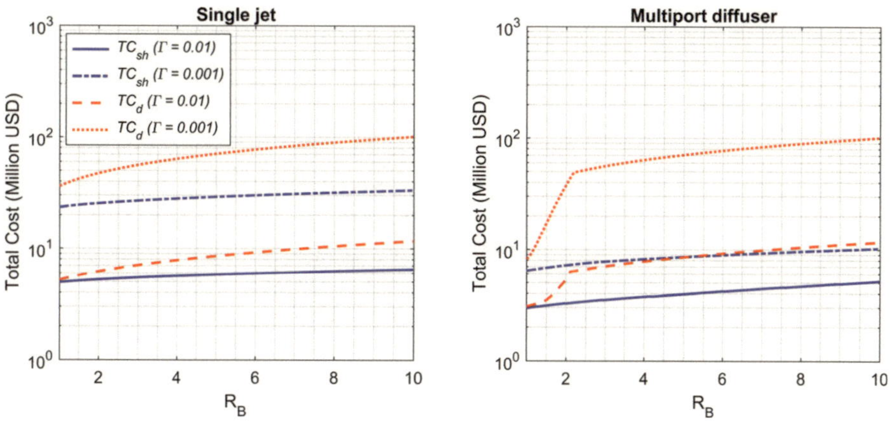

Figure 9.
Comparison of total costs at locations with $\Gamma = 0.01$ and $\Gamma = 0.001$ for the discharge of brine pre-diluted with SW using a single jet and a tee diffuser with $Q_b = 1\ m^3/s$ and $\Delta s_{th} = 2$ ppt.

Design	Variables	$\Gamma = 0.01$	$\Gamma = 0.001$
Single jet with submerged plume	H (m)	10.1	10.1
	u_0 (m/s)	5.4	5.4
	TC (Million USD)	5.2	35.9
Single jet with surfacing plume	H (m)	8.2	5.2
	u_0 (m/s)	8.3	20.8
	TC (Million USD)	4.9	23.3
Unidirectional diffuser with submerged plume	H (m)	2.8	1.4
	u_0 (m/s)	2.8	2.0
	N	26	150
	TC (Million USD)	3.1	8.2
Unidirectional diffuser with surfacing plume	H (m)	2.2	0.8
	u_0 (m/s)	4.8	8.2
	N	29	174
	TC (Million USD)	3.0	6.4

Table 5.
Example showing calculated design variables for discharge of brine (without pre-dilution or pre-concentration) at two locations with $\Gamma = 0.01$ *and* $\Gamma = 0.001$.

TWE and CW; since these effluents have to be discharged anyway), which will likely be more than the cost of discharging the blended effluent.

Unlike TWE and CW, SW does not need a separate outfall. In fact, intake of seawater for pre-dilution adds an extra cost. Also, as shown in **Figure 7**, the total cost increases with increase in R_B for the case of pre-dilution with SW. Thus, pre-diluting brine with SW is not economical. But it might be needed if there are regulatory restrictions on discharge concentrations themselves which are not met without pre-dilution.

For the calculations in this paper, a wide range of R_B (1 to 10) is considered. The flow rate of condenser cooling water from power plants is usually quite high as compared to the flow rate of brine. Therefore, a high value of R_B is possible for CW. However, the availability of TWE for blending with brine can be limited as it can be re-used or used for other purposes (e.g., irrigation).

8. Conclusions

Brine management strategies cause changes to the discharge flow rate, discharge concentrations of contaminants and the density difference between the effluent and seawater, and thus require changes to the outfall design. It is shown that pre-dilution with seawater is less economical than the discharge of brine without any pre-dilution. Thus, seawater should only be used for pre-dilution if there are restrictions on discharge concentrations of contaminants and other effluents (TWE or CW) are not available for pre-dilution. Concentration of brine is also not viable from an environmental standpoint. On the other hand, pre-dilution with TWE or CW is likely to be economically beneficial.

For the design of a new outfall for a desalination plant with known amount of pre-dilution or pre-concentration, design variables are calculated for both a single port and a multiport outfall. Depending on the environmental regulations which might have restrictions on plume visibility, design parameters are evaluated for a submerged plume or a surfacing plume. It is shown that when the plume is allowed to hit the water surface (no restrictions on plume visibility), the required water depth and total cost of the outfall can be significantly reduced. For such cases, the required water depth and the offshore distance decrease as the blending ratio increases. At locations which require the plume to be submerged, the design with a single jet is found to have lower cost than a design with multiple ports (for most values of the blending ratio). However, for locations with no restrictions on plume visibility, use of a multiport diffuser is recommended as it can result in much lower cost than a single jet.

The effect of bottom slope and threshold concentrations on outfall design is also explored. Locations with mild bottom slope encourage the use of outfalls with multiple ports which can reduce the required water depth and, in turn, the offshore distance of the outfall from the shoreline. An increase in threshold concentrations usually leads to a reduction in outfall cost as the outfall needs to achieve a smaller dilution. Similarly, more stringent regulations (smaller threshold concentrations) can lead to a rapid increase in outfall cost.

Acknowledgements

This work was supported by Kuwait-MIT Center for Natural Resources and the Environment (CNRE), which was funded by Kuwait Foundation for the Advancement of Sciences (KFAS).

Conflict of interest

The authors declare no conflict of interest.

Author details

Ishita Shrivastava* and Edward Eric Adams
Massachusetts Institute of Technology, Cambridge, MA, USA

*Address all correspondence to: ishita@mit.edu

References

[1] Bleninger T, Jirka GH. Environmental planning, Prediction and Management of Brine Discharges from Desalination Plants. Muscat, Sultanate of Oman: Middle East Desalination Research Center (MEDRC), 2010.

[2] Chow AC, Verbruggen W, Morelissen R, Al-Osairi Y, Ponnumani P, Lababidi HMS, Al-Anzi B, Adams EE. Numerical prediction of background buildup of salinity due to desalination brine discharges into the northern Arabian gulf. Water. 2019: 11: 2284. DOI:10.3390/w11112284

[3] Chung HW, Nayar KG, Swaminathan J, Chehayeb KM, Lienhard V JH. Thermodynamic analysis of brine management methods: Zero-discharge desalination and salinity gradient power production. Desalination. 2017: 404: 291-303. DOI: 10.1016/j.desal.2016.11.022

[4] Choi S, Kim B, Nayar KG, Yoon J, Al-Hammadi S, Lienhard V JH, Han J, Al-Anzi B. Techno-economic analysis of ion concentration polarization desalination for high salinity desalination applications. Water Res. 2019: 155: 162-174. DOI:10.1016/j.watres.2019.02.023

[5] Akram W, Sharqawy MH, Lienhard V JH. Energy utilization of brine from an MSF desalination plant by pressure retarded osmosis. Proceedings of the IDA world congress on desalination; Oct. 20-25 2013; Tianjin, China.

[6] Weiner AM, McGovern RK, Lienhard V JH. Increasing the power density and reducing the levelized cost of electricity of a reverse electrodialysis stack through blending. Desalination. 2015: 369: 140-148. DOI:10.1016/j.desal.2015.04.031

[7] Mei Y, Tang CY. Recent developments and future perspectives of reverse electrodialysis technology: A review. Desalination. 2018: 425: 156-174. DOI:10.1016/j.desal.2017.10.021

[8] Strathmann H. Electrodialysis, a mature technology with a multitude of new applications. Desalination. 2010: 264: 268-288. DOI:10.1016/j.desal.2010.04.069

[9] Kim SJ, Ko SH, Kang KH, Han J. Direct seawater desalination by ion concentration polarization. Nat Nanotech. 2010: 5: 297-301. DOI: 10.1038/nnano.2010.34

[10] Al-Mutaz IS, Al-Namlah AM. Characteristics of dual purpose MSF desalination plants. Desalination. 2004: 166: 287-294. DOI:10.1016/j.desal.2004.06.083

[11] Shrivastava I, Adams EE. Effect of shallowness on dilution of unidirectional diffusers. ASCE J Hydraul Eng. 2019: 145(12): 06019013. DOI: 10.1061/(ASCE)HY.1943-7900.0001640

[12] Environmental, Health and Safety guidelines for water and sanitation. International Finance Corporation, World Bank Group, 2007.

[13] Abessi O, Roberts PJW. Dense Jet Discharges in Shallow Water. ASCE J Hydraul Eng. 2015: 142 (1): 04015033. DOI:10.1061/(ASCE)HY.1943-7900.0001057

[14] Roberts PJW, Ferrier A, Daviero G. Mixing in inclined dense jets. ASCE J Hydraul Eng. 1997: 123 (8): 693–699. DOI:10.1061/(ASCE)0733-9429(1997)123:8(693)

[15] Shao D, Law AWK. Mixing and boundary interactions of 30° and 45° inclined dense jets. Environ Fluid Mech. 2010: 10 (5): 521-553. DOI:10.1007/s10652-010-9171-2

[16] Lai CCK, Lee JHW. Mixing of inclined dense jet in stationary ambient. J Hydro-environ Res. 2012: 6(1): 9-28. DOI:10.1016/j.jher.2011.08.003

[17] Jiang B, Law AWK, Lee JHW. Mixing of 30° and 45° inclined dense jets in shallow coastal waters. ASCE J Hydraul Eng. 2014: 140 (3): 241-253. DOI:10.1061/(ASCE)HY.1943-7900.0000819

[18] Adams EE. Dilution analysis for unidirectional diffusers. ASCE J Hydraul Div. 1982: 108 (HY3): 327-342. DOI: 10.1061/JYCEAJ.0005833

[19] Lai ACH, Lee JHW. Dynamic interactions of multiple buoyant jets. J Fluid Mech. 2012: 708: 539-575. DOI: 10.1017/jfm.2012.332

[20] Abessi O, Roberts PJW. Multiport Diffusers for Dense Discharges. ASCE J Hydraul Eng. 2014: 140 (8), 04014032. DOI:10.1061/(ASCE) HY.1943-7900.0000882

[21] Adams EE, Skamarock WC, Nothaft RU. Shore line effects on mixing of tee diffuser. ASCE J Hydraul Div. 1982: 108 (HY10): 1232-1238. DOI: 10.1061/JYCEAJ.0005921

[22] Shrivastava I, Adams EE. Mixing of tee diffusers in shallow water with crossflow: A new look. ASCE J Hydraul Eng. 2019: 145(4): 04019006. DOI:10.1061/(ASCE) HY.1943-7900.0001574

[23] Jiang B, Law AWK. Non-interfering multiport brine diffusers in shallow coastal waters. J Appl Water Eng and Res. 2013: 1(2): 148-157. DOI:10.1080/23249676.2013.878883

[24] Maalouf S, Rosso D, Yeh WWG. Optimal planning and design of seawater RO brine outfalls under environmental uncertainty. Desalination. 2014: 333: 134-145. DOI: 10.1016/j.desal.2013.11.015

[25] Chang NB, Yeh SC, Chang CH. Optimal expansion of a coastal wastewater treatment and ocean outfall system under uncertainty (II): Optimisation analysis. Civ Eng and Environ Syst. 2011: 28(1): 39-59. DOI: 10.1080/10286600903243138

[26] Shrivastava I, Adams EE. Pre-dilution of desalination reject brine: Impact on outfall dilution in different water depths. J Hydro-Environ Res. 2019: 24, 28-35. DOI:10.1016/j.jher.2018.09.001

[27] Lattemann S, Höpner T. Environmental impact and impact assessment of seawater desalination. Desalination. 2008: 220: 1-15. DOI: 10.1016/j.desal.2007.03.009

[28] Roberts PJW, Salas HJ, Reiff FM, Libhaber M, Labbe A, Thomson JC. Marine Wastewater Outfalls and Treatment Systems. London (UK): IWA Publishing; 2010. DOI:10.2166/9781780401669

[29] West Basin Municipal Water District Ocean Water Desalination Program Master Plan (PMP), Volume II. Irvine (CA): Malcolm Pirnie, 2013.

[30] Camp Pendleton Seawater Desalination Project Feasibility Study, Prepared for San Diego County Water Authority, Volume I. San Diego (CA): RBF Consulting, 2009.

[31] Fischer HB, List EJ, Imberger J, Koh RCY, Brooks NH. Mixing in Inland and Coastal Waters. San Diego (CA): Academic Press; 1979. DOI:10.1016/C2009-0-22051-4

[32] Grace RA. Marine Outfall Construction. Reston (VA): ASCE Press; 2009. DOI:10.1061/9780784409848

[33] Davis CV. Handbook of Applied Hydraulics. New York (NY) and London (UK): McGraw-Hill; 1942.

[34] Desalination for Safe Water Supply, Guidance for the Health and Environmental Aspects Applicable to Desalination. Public health and the environment. World Health Organization, 2007.

[35] General EHS Guidelines: Wastewater and ambient water quality. International finance corporation, World Bank Group, 2007.

[36] State Water Resources Control Board (SWRCB), California Environmental Protection Agency. 2015. "Water Quality Control Plan, Ocean Waters of California (Ocean Plan)."

[37] Almquist CW, Stolzenbach KD. Staged multiport diffusers. ASCE J Hydraul Div. 1980: 106 (HY2): 285-302. DOI:10.1061/JYCEAJ.0005366

Section 3

Water Utilization, Remineralization and Acid Control

Chapter 4

Desalination and Agriculture

Issam Daghari

Abstract

In arid countries like Tunisia, the need to find new sources of water for irrigation has become imminent. Desalination of seawater can be an alternative to irrigation. Water desalination is a process that makes it possible to obtain freshwater (drinking water or, more rarely, due to the cost, usable for irrigation) from brackish or saltwater (seawater in particular). In this article, we take a look at the leading food companies specializing in desalination of irrigation around the world and the prospects for the solar energy desalination potential for irrigation in Tunisia. We have noticed that several companies invest money to desalinate water for agricultural purposes. However, the cost of a cubic meter of water sometimes remains high to go forward with this new technology.

Keywords: irrigation, renewable energy, solar desalination, water cost

1. Introduction

Since the 1960s, water treatment and water desalination have been the subject of scientific studies in the USA [1]. Among these studies, water specialists have tried to see the performance of thermal desalination plants with thermodynamic cycle plants to have high energy efficiency, large solar plants called concentrating solar power (CSP) can thus be associated with multi-effect distillation (MED) or multi-stage flash distillation (MSF) desalination stations [2]. These investigations verified the technical feasibility of setting up desalination stations for agricultural purposes. These studies have shown that an irrigated perimeter can have a water supply through desalination plants for sustainable agricultural production. This can help ensure competitive prices on the market for agricultural products [3]. Several successful experiences exist in this area.

In the Bay Lagoon in the Philippines [4], the main challenge for agriculture is the salinity of the water, which exceeds 2 g/L. The establishment of a desalination plant for irrigation has enabled the increase in agricultural production for the water supply of 30,000 ha. In this sense, several countries have chosen desalination for water supply. Arid countries such as Tunisia, Saudi Arabia, and Egypt, and less arid countries such as Nigeria, China, Indonesia, and Cuba have set up drinking water desalination plants [5]. Half of Malta's municipal water quantity comes from seawater desalination [6]. In Kuwait, there have been years of excess desalinated water over requirements [7].

Nowadays, desalination is a method to obtain good quality water. All desalination processes are energy intensive and share the common minimum energy required to cause the saline solution to separate into pure water and concentrated brine. It is dependent on the detailed technology used, the exact mechanism, or the number of process steps. Furthermore, the overall equivalent power consumption of a multi-stage flash (MSF) unit is 20–30 kWhelec/m^3, the overall equivalent power consumption of the multi-effect desalination (MED) unit also varies from 15 to 22 kWhelec/m^3 [8, 9]. These thermal processes are energy intensive because there is a loss of energy efficiency due to phase changes (fossil to thermal or fossil to electric to thermal). According to the same author [10], for the reverse osmosis (RO) membrane process, the overall equivalent power consumption of the SWRO unit (seawater RO) reached the lowest specific energy consumption level of SWRO at 2.00 kWhelec/m^3. The energy consumed by conventional desalination plants usually comes from the combined cycle power plants. They are characterized by the highest efficiency of electricity generation technology from fossil fuels. These units are among the most developed, currently achieving yields above 60%. The CO_2 emission is equal to 330 kg—CO_2/MWh [11]. Speaking of agricultural water consumption, according to the United Nations report [12], agriculture alone uses 70% of the world's water supply. Besides, global food demand is expected to increase by another 70% by 2050. However, according to the report, the main challenge facing the world today is not so much the increase of food production, but rather to provide good-quality irrigation water to farmers in sufficient quantities. The shortage of water in arid zones has led to the usage of low-quality irrigation water in agriculture in most arid climate areas [13]. The water deficit is a problem present in many parts of the world, with lower rainfall and increased salinity of aquifers [14].

There is another method of desalination that finds success among manufacturers. This is "MEDAD" desalination which is a hybrid of traditional multi-effect distillation (MED) and adsorption cycle (AD) [15]. In general, there are several hybridization trends. In RO processes, intake, pretreatment, and brine disposal cost 25% of total desalination cost at 30–35% recovery. Shahzad and Ng [16] proposed a tri-hybrid system to enhance overall recovery up to 81%. The conditioned brine leaving from RO processes is supplied to proposed multi-evaporator adsorption cycle driven by low-temperature industrial waste heat sources or solar energy. The brine rejection concentration of the tri-hybrid cycle may vary from 166,000 to 222,000 ppm if the concentration of RO retentate varies from 45,000 to 60,000 ppm.

Several coastal countries see water desalination as a solution to water scarcity [17]. In Algeria, a neighboring country of Tunisia, desalination is considered by water experts as the only solution present to avoid a future water shortage [18]. The very specific conditions of the Mediterranean Sea (freshwater at 19°C and a salinity of 38 g/L [19], while the waters of the golf course are at 30°C and a salinity of 40.5 g/L [20]) will result in lower cost per cubic meter of desalinated water Algeria started building large-scale desalination plants after the 2001 water crisis, with a total capacity of over 2 million m^3/d [21].

Also in Libya, which is Tunisia's second neighbor country, a serious effort has been made to develop additional water sources from desalination [22] and the country has about 10 desalination stations.

There are great advances in the field of membrane and thermal desalination in particular the specific energy consumption (**Table 1**). According to Shahzad et al. [8], the performance of desalination plants, conventionally reported based on fossil fuels, can now be transformed fairly on a common platform based on specific energy consumption (SPE). This new factor, called the standard universal performance ratio (SUPR), is calculated on the basis of SPE and presented in **Table 1**. Thermal processes have better efficiency of 2.82 and 2.00 m^3/kWh for

Specific energy consumption and performance ratio	Reverse osmosis (SWRO)	Multi-stage flashing (MSF)	Multi-effect distillation (MED)
Electricity (kWh_{elec}/m^3) [23, 24]	3.54	2.82	2.00
Thermal (kWh_{ther}/m^3) [23, 24]	—	90.0	70.0
Equivalent standard primary energy (SPE) and standard universal performance ratio (SUPR)			
Conversion factor for electricity (weighted CF_{elec})	2.012		
Conversion factor for thermal for less than 130°C operation (CF_{ther})	—	35.33	
Standard primary energy (Q_{SPE})	7.12	8.22	6.00
$Q_{SPE} = \left[\left(kWh_{elec} / m^3\right)\left(CF_{elec}\right)\right] + \left[\left(kWh_{ther} / m^3\right)\left(CF_{ther}\right)\right]$			
Standard universal performance ratio (SUPR)	90.7	78.6	107.6
$SUPR = 2326 / \left(3.6 Q_{SPE}\right)$			
SUPR % of thermodynamic limit (SUPR = 828 at TL)	10.9%	9.5%	13.0%

Table 1.
SPE and SUPR calculation of major desalination processes.

MSF and MED respectively with respect to RO. The CSP + MSF or MED configuration should experience a boom in the future.

2. Emerging agro-industrial companies specializing in solar desalination for irrigation

Sundrop Farms is a leading company in high value-added horticulture. It is developing irrigated perimeters in arid areas of Australia with desalinated water from solar energy as the water source [25]. It is an agri-food company with the technological know-how to develop and operate hydroponic greenhouses (**Figure 1**).

Desalination is a long-term solution. But the high energy requirements of desalination are a drawback [27]. This is why Sundrops farms use solar desalination. The technology has given satisfactory technical results. In 2015, Sundrop Farms built a 20-hectare solar greenhouse in Port Augusta. A concentrated solar power plant (CSP) desalinates seawater taken from the Spencer Gulf to irrigate agricultural products [28].

Figure 1.
An integrated desalination facility drawing its main electricity from an adjacent concentrating solar power plant [26].

Tomato yields reached 850 tonnes/ha in hydroponic greenhouses [24]. A quantity of desalinated water of 335,103 m^3 is used in the greenhouses. The financial cost of the project is estimated at the US $ 205 million [29].

In the same context, a study carried out in 2017 in Spain showed that the supply of drinking water and irrigation water *via* a desalination plant increased resilience in the face of water shortages [30].

Another similar project was built in 2012 as part of the Norwegian agro-industry called the Sahara Forest Project in Qatar on an area of 300 ha. The Qatari installation can provide desalinated water for all crop irrigation needs [31] for a yield of 633 tonnes/ha of tomatoes and melons. This same agro-industrial project in the Sahara forest will initiate numerous activities in Tunisia over 10 ha [26]. In addition, an American blog called "sustainable business" specializing in sustainable development projects described a solar thermal project launched in 2014 in California [23].

A Californian start-up called WaterFX has set up a 14-hectare solar thermal desalination plant. It supplied 3.8 million m^3 of water in 2014 from the saline drainage water of the San Joaquin Valley and turns it into freshwater for irrigation. The desalination unit recycles drainage water over an area of 2800 ha into a source of freshwater for nearby irrigated areas.

The success of the project convinced the Panoche Water District, in an arid agricultural region of California, to build a desalination plant [32]. WaterFX's operating cost is the US $ 450 for 1.2103 m^3 [33]. The price was deemed acceptable by the Panoche Water District.

A current price of US $ 1/m^3 is considered acceptable in some countries for domestic and industrial uses [34].

3. Solar desalination in Tunisia: perspectives

Several coastal countries see water desalination as a solution to water scarcity [17]. In Algeria, a neighboring country of Tunisia, desalination is considered by water experts as the only solution present to avoid a future water shortage [18]. The very specific conditions of the Mediterranean Sea (freshwater at 19°C and a salinity of 38 g/L [19], while the waters of the golf course are at 30°C and a salinity of 40.5 g/L [20]) will result in lower cost per cubic meter of desalinated water Algeria started building large-scale desalination plants after the 2001 water crisis, with a total capacity of over 2 million m^3/d [21].

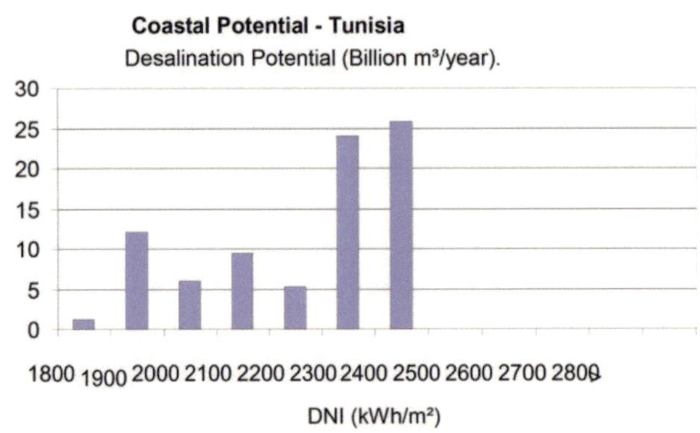

Figure 2.
The statistical analysis of the direct normal irradiance (DNI) map for CSP-desalination in Tunisia (20 m above sea level) [35].

Also in Libya, which is Tunisia's second neighbor country, a serious effort has been made to develop additional water sources from desalination [22] and the country has about 10 desalination stations. The German Institute for Space Research carried out a study on the problem of water scarcity in the Arab region.

The main conclusion is that this shortage can be alleviated by resorting to water desalination by solar energy. Solar desalination can provide more than 85 billion m^3/year in some areas when the direct normal irradiance (DNI) exceeds 1800 kWh/m^2/year (**Figure 2**). The x-axis represents the DNI, and the y-axis represents the quantity of desalinated water obtained (in billions of m^3/year).

This is the distribution of surfaces according to the radiation expressed in the production of desalinated water. Considering the increase in population in most countries of the South and particularly those which are arid like Tunisia, the desalination of seawater will be an important source of water in the water supply [36]. In addition, with the development of seawater desalination and the expected progress in terms of access to drinking water and sanitation (United Nations Sustainable Development Goal SDG 6), this sector could reduce its consumption. In energy by 15% by 2040 [37] everywhere, when political decision-makers wanted it, deserts were irrigated in several cities of the Middle East with desalinated seawater [38].

4. Major challenges for the construction of solar desalination plants for irrigation

The launch of desalination plants presents many major challenges. A well-detailed expertise file must be drawn up on the technical requirements [39]. There is a need to collect a lot of data for the construction of these stations. These data concern the distribution of water and its use for irrigation for better agricultural efficiency. Also, the data concern resources, both in their quantitative and qualitative aspects (rain, infiltration, flows, recharge, and water quality). This idea is shared by Margat [40] who says that desalination operations bring together a complex chain of water development and control for better energy efficiency. Indeed, when the occupation of the coast poses a problem for the installation of desalination units, the problem is "energy efficiency," which must be fully taken into account for these units coupled with thermal power stations [41]. The second major challenge is that the cost per cubic meter of water is expensive in some areas [42]. Indeed, the desalination of certain brackish water requires a fairly significant pretreatment, which can affect this cost. Since reverse osmosis desalination plants can be set up in isolated sites, another existing challenge is that these stations could significantly affect marine life and fauna [43] leading to ecological variations. For example, thermal discharges can seriously harm the marine ecosystem.

There are also steps to be taken to change consumers' perception of desalinated water, which will enable the supply of sufficient quantities of water and in fact constitute new water insurance [44]. In Tunisia, water desalination is technically feasible but at excessively high costs [45]. Indeed, there is a transfer of water from dams that mixes in aquifers to fight against seawater intrusion in coastal regions and not with desalinated water [46].

5. Conclusion

Agri-food companies are emerging in countries suffering from increasing water stress. The goal is to develop agricultural production through solar water desalination. This will be one of the alternatives on which Tunisia can count for the supply of irrigation water. The economy will flourish with the irrigation water needed for millions of hectares of arable land. However, the first and main concern is the cost of these factories that pose a problem for poor or emerging countries. Second, emissions of gas, hot water, and salinity create environmental problems. Third, using chlorine to clean membranes (reverse osmosis) creates chemical water that cannot be discharged into the sea.

Author details

Issam Daghari
National Institute of Agronomy, University of Carthage, Tunis, Tunisia

*Address all correspondence to: issam.daghari@gmail.com

References

[1] Gomella C. Diffusion de l'ozone dans l'eau. La Houille Blanche. 1967;**4**:423-429

[2] Thiennot R. Énergie thermique des mers: Les centrales à cycle thermodynamique fermé. La Houille Blanche. 1981;**4**(5):323-330

[3] Picard J. Les aspects socio-économiques de la valeur de l'eau. La Houille Blanche. 1977;**2**(3):237-241

[4] Pariset E. Laguna de Bay, Philippines: Étude de la protection et de l'utilisation des ressources en eaux. La Houille Blanche. 1977;**2**(3):201-209

[5] André B. Le tuyau bonna: un siècle de présence à l'étranger. La Houille Blanche. 1995;**4**:98-102

[6] Margat J. De l'adéquation des ressources, tant en quantité qu'en qualité. La Houille Blanche. 1993;**6**(7):215-224

[7] Detay M, Bersillon J-L. La réalimentation artificielle des nappes profondes: faisabilité et conséquences. La Houille Blanche. 1996;**4**:57-61

[8] Shahzad MW, Burhan M, Ang L, Choon Ng K. A standard primary energy approach for comparing desalination processes. npj Clean Water. 2019;**1**:1-7

[9] Shahzad MW, Burhan M, Hyuk Soo S, Seung Jin O, Choon Ng K. Desalination processes evaluation at common platform: A universal performance ratio (UPR) method. Applied Thermal Engineering. 2018;**134**:62-67

[10] Shahzad MW, Burhan M, Ang L, Choon Ng K. Energy-water-environment nexus underpinning future desalination sustainability. Desalination. 2017;**413**:52-64

[11] Kotowicz J, Brzęczek M. Analysis of increasing efficiency of modern combined cycle power plant: A case study. Energy. 2018;**153**(C):90-99

[12] United Nations. Water for Food (UNCTAD). 2011. n°4

[13] Hamdi Y. Frequency analysis of droughts using historical information—New approach for probability plotting position: Exceedance probability. International Journal of Global Warming. 2011;**3**(1/2)

[14] Bahir M, Chkir N, Trabelsi R, Friha HA, Zouari K, Chammati H. Hydro-geochemical behaviour of two coastal aquifers under severe climatic and human constraints: Comparative study between Essaouira basin in Morocco and Jeffara basin in Tunisia. International Journal of Hydrology Science and Technology. 2012;**2**(1):75-100

[15] Shahzad MW, Ng KC, Thu K, Saha BB, Chun WG. Multi effect desalination and adsorption desalination (MEDAD): A hybrid desalination method. Applied Thermal Engineering. 2014;**72**:289-297

[16] Shahzad MW, Ng KC. An improved multi-evaporator adsorption desalination cycle for GCC countries. Energy Technology. 2017. DOI: 10.1002/ente.201700061

[17] Wonham J. Ocean cities: Environmental aspects. La Houille Blanche. 1995;**8**:60-62

[18] Bessenasse M. Evaluation du coût des produits chimiques utilisés dans trois stations de dessalement du littoral algérois. La Houille Blanche. 2009;**3**: 138-147

[19] Skliris N, Zika JD, Herold L, Josey SA, Marsh R. Mediterranean Sea water budget long-term trend inferred from salinity observations. Climate Dynamics. 2018;**51**:2857-2876. DOI: 10.1007/s00382-017-4053-7

[20] Ibrahim HD, Xue P, Eltahir EAB. Multiple salinity equilibria and resilience of Persian/Arabian Gulf Basin salinity to brine discharge. Frontiers in Marine Science. 2020. DOI: 10.3389/fmars.2020.00573

[21] Verdier J, Viollet P-L. Les tensions sur l'eau en Europe et dans le bassin méditerranéen: Des crises de l'eau d'ici 2050. La Houille Blanche. 2015;**6**: 102-107

[22] EI-Gheriani AM. The great man-made river project. La Houille Blanche. 2003;**1**:99-101

[23] Sustainable Business. First solar desalination plant will serve California farmers. 2014. Available from: http://www.sustainablebusi ness.com/first-solar-desalination-plant-will-serve-california-farm ers-52988/

[24] The Guardian. Water-smart farming: How hydroponics and drip irrigation are feeding Australia. 2017. Available from: https://www.theguardian.com/sustainable-business/2017/apr/27/water-smart-farming-how-hydroponics-and-drip-irrigation-are-feeding-australia

[25] Cleantechnica. Can solar thermal desalination make sustainable agriculture possible? 2014. Available from: https://clean technica.com/2014/12/05/can-solar-thermal-desalination-make-sustainable-agriculture-possible/

[26] Inhabitat. Australian desert farm grows 17,000 metric tons of vegetables with just seawater and sun. 2016. Available from: http://inhabitat.com/australian-desert-farm-grows-17000-metric-tons-of-vegetables-with-just-seawater-and-sun/

[27] Lafforgue M. Supplying water to a water-stressed city: Lessons from Windhoek. La Houille Blanche. 2016;**4**: 40-47

[28] Council Development Assessment Panel Agenda Meeting #123. Port Augusta City Council. 2014 [Accessed: 12 August 2014]

[29] Government of South Australia. Focus—South Australian Major Developments. 2015. p. 44

[30] Barraque B, Isnard L, Souriau J. European urban water crisis: The management dimension. La Houille Blanche. 2017;**2**:27-34

[31] Renew Economy. Solar and seawater turn desert into a greenhouse. 2013. Available from: http://reneweconomy.com.au/solar-and-seawater-turn-desert-into-a-greenhouse-28833/

[32] Earthtechling. Solar desalination to tramp up in California. EarthTechling. 2015. Available from: http://earthtechling.com/2015/08/solar-desalination-california/

[33] Forbes. WaterFX sees solar desalination as one way to address the world's water problem. 2014. Available from: http://www.forbes.com/

[34] Babillot P, Le Lourd P. Is there a water market? La Houille Blanche. 2000;**2**:39-54

[35] Fichtner. Desalination using renewable energy part II final report. MENA Regional Water Outlook. 2011

[36] La Société Hydrotechnique de France. De l'avenir en général et de l'eau en particulier. La Houille Blanche. 1998;**2**:25-26

[37] Loudière D, Gourbesville P. Rapport mondial des Nations Unies sur la mise en valeur des ressources en eau 2020: L'eau et les changements climatiques. La Houille Blanche. 2020;**3**:76-81

[38] Dunglas J. La technique peut-elle tout résoudre? La Houille Blanche. 2000;**2**:29-38

[39] Ghaffour N, Missimer T, Amy G. Technical review and evaluation of the economics of water desalination: Current and future challenges for better water supply sustainability. Desalination. 2013;**309**:197-207. DOI: 10.1016/j.desal.2012.10.015

[40] Margat J. Combien d'eau utilise-t-on et use-t-on? Pour quoi faire? La Houille Blanche. 2000;**2**:12-28

[41] Esmaeilion F. Hybrid renewable energy systems for desalination. Applied Water Science. 2020;**10**:84. DOI: 10.1007/s13201-020-1168-5

[42] Wiesman R. IDA desalination inventory. Desalination & Water Reuse Quarterly. 2003;**12**(3):10-13

[43] Miri R, Chouikhi A. Ecotoxicological marine impacts from seawater desalination plants. Desalination. 2005;**182**(1-3):403-410. DOI: 10.1016/j.desal.2005.02.034

[44] Cabrera E, Estrela T, Lora J. Desalination in Spain. Past, present and future. La Houille Blanche. 2019;**1**:85-92

[45] Daghari I, Zarroug MR. Concepts review of solar desalination technologies for irrigation: Bibliographic review. Journal of New Sciences: Agriculture and Biotechnology. 2020;**71**(3): 4319-4326

[46] Daghari I, El Zarroug M-R, Muanda C, Shanak N. Best irrigation scheduling way with saline water and desalinated water: Field experiments. La Houille Blanche. 2020;**4**:72-74

Chapter 5

Remineralization and Stabilization of Desalinated Water

Nicholas Nelson and Antonella De Luca

Abstract

Permeate or distillate from desalination processes is typically void of minerals and alkalinity, inherently acidic and therefore corrosive to water distribution infrastructure. The reintroduction of both minerals and alkalinity is essential for the stabilization of the water before it is sent to consumers making this the last step of the treatment process. Classical water stability is evaluated with respect to its calco-carbonic equilibrium which looks at the balance of calcium hardness, alkalinity and pH to determine whether the water has a tendency to dissolve or precipitate calcium carbonate. The purpose of remineralization processes is replenish the levels of calcium hardness and alkalinity in the water and then adjust the pH to deliver a stable water quality that is safe for human consumption and non-aggressive to water distribution infrastructure.

Keywords: remineralization, stabilization, potabilization, post-treatment, hardness, calcium, magnesium, alkalinity, bicarbonate, pH, corrosion, calcium carbonate, calcite contactors, lime, saturator, carbon dioxide, chemicals, indices

1. Introduction

As the name suggests, desalination processes are responsible for the removal of dissolved salts or ions from sea-, brackish- or tertiary-treated wastewater. These processes however, are not selective, and desirable minerals are just as readily removed along with the unwanted salts. The resultant desalination permeate or distillate is therefore void of essential minerals and alkalinity, rendering it unstable and corrosive. If left untreated, desalinated water will corrode the distribution infrastructure and will deteriorate the linings used as a protective barrier between the conduit and the drinking water. This creates a huge financial burden on asset owners. Recent estimates by the American Water Works Association suggested a program in excess of $300 billion over a 20-year period would need to be spent in the US alone to replace pipe damaged as a result of corrosion [1]. The financial impact of corrosion does not stop with the cost of pipework replacement either: a recent study by the Chicago-based Center for Neighborhood Technology in conjunction with the Chicago Metropolitan Agency for Planning found that up to 22 billion gallons of potable water is lost on a yearly basis due to leaks within the distribution network [2].

More important than the cost of pipework replacement however, is the health risk posed to consumers of non-stabilized water. Corrosion of pipework by aggressive water can lead to the release of heavy metal ions such as lead, copper and iron

IntechOpen

to levels that are considered unsafe for human consumption, inducing "red water" incidents [3] and posing health risks to its consumers. Lead, for example, is considered one of the most toxic heavy metals and has been shown to impede both the physical and mental developments in children [4]. Ingestion of copper on the other hand, can result in liver and kidney damage. Cadmium, found in galvanized piping, is also highly toxic, considered a carcinogenic and even short-term exposure can lead to problems of the liver, heart and kidney [5]. Finally trace metals such as cadmium, barium, chromium and aluminum have been found to leach from the mortar linings of concrete lined pipes [6]. Therefore it is critical that following desalination processes, minerals and alkalinity are added to the water to achieve a buffered and stabilized water quality that is non-corrosive and safe for use.

2. Targets for remineralization

When designing remineralization systems and determining the appropriate water quality targets, a handful of key parameters are taken into consideration: alkalinity, calcium hardness and pH. The importance of these parameters are discussed further in detail in the following sections.

2.1 Importance of alkalinity

Alkalinity is one of the most important water quality parameter due to its ability to maintain a stable and buffered system, as well as its effectiveness in protecting against various mechanisms of corrosion. In general terms, alkalinity is defined as the capacity of an aqueous solution to accept a proton, or rather to neutralize an acid. Although other systems can contribute to alkalinity, for drinking water processes, it is only the carbonate system that is taken into account to define alkalinity. This is due to its predominance over other buffering systems in the pH ranges usually associated with water chemistry. Alkalinity for drinking water is therefore expressed as the following:

$$\text{Alkalinity} = 2\left[CO_3^{2-}\right] + \left[HCO_3^{-}\right] + \left[OH^{-}\right] - \left[H^{+}\right] \tag{1}$$

As seen in Eq. (1), alkalinity is determined primarily by the concentration of carbonate and bicarbonate ions, and to a lesser extent by the concentration of hydroxide and hydrogen ions, which also define the pH. Bicarbonate ions are particularly important, due to their ability to consume both hydrogen and hydroxide ions, and therefore provide a buffer to pH shifts in both directions.

In addition to providing a buffer to protect against shifts in pH, alkalinity is one of the most important parameters in corrosion control. The World Health Organization recommends high alkalinity levels as a suitable technique for preventing many mechanisms of corrosion in their Guidelines to Drinking Water Quality [7]. This is backed up by a multitude of studies into corrosion control and experts responsible for setting water quality targets.

More particularly, alkalinity is essential in controlling the corrosion of many metal materials of construction. One of the best methods of controlling iron, copper, zinc or galvanized iron corrosion, is the precipitation of the respective carbonates, such as siderite ($FeCO_3$), basic copper carbonate and basic zinc carbonate, for the formation of a passivation layer on the surface of the material [8]. A higher alkalinity content is therefore imperative to ensure sufficient carbonate species are available for the formation of these compounds. These protective layers are also an effective strategy against microbiologically induced pitting [9]. Lead on the other hand, is released into the water either directly from the pipe, or from lead-containing corrosion products that are formed on the pipe surface. In terms of lead corrosion products, the most common of these is lead carbonate whose propensity of lead (II) carbonates to dissolve into the water stream is directly related to the concentration of carbonate species already within the water. Finally, alkalinity is essential to prevent the degradation of concrete and cement-based systems that in many cases are used as a lining to protect large bore conduits constructed from mild steel or ductile iron.

2.2 Importance of calcium

Next to alkalinity, calcium hardness is the most important parameter for post-treatment process for a number of reasons. Firstly, it is the most suitable counter-ion to the anionic alkalinity species. Other potential alternatives can result in water that is toxic to plant life (e.g. sodium ions), or are not so readily available (e.g. potassium). Calcium carbonate on the other hand, is one of the most readily sourced minerals in the world, making up 4% of the earth's crust [10]. Secondly, it is the concentration of calcium ions in addition to alkalinity and pH that defines the calco-carbonic equilibrium of the water. The calco-carbonic equilibrium defines a water's propensity to dissolve or precipitate calcium carbonate and is the primary indication of water stability for a number of reasons. Water that is aggressive to calcium carbonate, will be aggressive to concrete or cement-lined pipe, along with asbestos cement pipe. Because of the amount of water distribution infrastructure that is either made from concrete (storage tanks) or cement-lined (mild-steel concrete-lined or ductile iron concrete-lined pipe), this represents one of the largest investment costs for utility owners and highlights the importance of protecting these from corrosion. Whilst on the other hand, the precipitation of a thin layer of calcium carbonate on the surfaces of water treatment infrastructure is considered a suitable strategy against corrosion for a wide range of materials.

Finally, calcium has a number of health benefits to the consumer, with increased calcium levels within drinking water being linked to decreased incidences of cardiovascular disease. This has been acknowledged and accepted by the World Health Organization, who have clearly stated in their background document for the development of Guidelines for Drinking Water Quality, that insufficient calcium intake is associated with increased risks of osteoporosis, kidney stones, hypertension, stroke, coronary artery disease, some cancers and even obesity [11]. Calcium in drinking water is not just important for the body, but also for the teeth. Demineralization and remineralization processes are constantly taking place on the surface of tooth enamel based largely on the surrounding fluid. Saturation of the surrounding fluid with respect to calcium is critical to promote the remineralization or repair of dental tissue [12].

2.3 Importance of pH

pH is the final parameter that is considered for determining quality as part of post treatment processes. The desired target pH is often determined as a by-product of the previously mentioned parameters: alkalinity and calcium content. This is due to the fact that 1) the dissolution or precipitation potential of the water with respect to calcium carbonate is often used as the key indicator to the stability of water; and 2) the alkalinity, calcium content and pH are the three major contributing factors in determining dissolution or precipitation potential of the water. As noted earlier, this calco-carbonic balance is extremely important for cement-lined or concrete infrastructure. In such circumstances, system designers often opt for a slightly precipitative water ($pH > pH_{sat}$) so that a protective scale is built up on the surface of the cement lined pipes or tanks. For other materials of construction, pH also appears to play an important role in controlling corrosion. The World Health Organization for example recommends a pH range of 6.8–7.3 to deal with iron corrosion, 8.0–8.5 to deal with lead and copper corrosion and a pH of less than 8.3 to deal with brass corrosion [13].

3. Methods for measuring and characterizing stabilized water

Individual parameters are not sufficient enough to predict the corrosivity of water and a number of indices have therefore been developed that look at the relationship of these parameters in an attempt to quantify the corrosivity or aggressivity of the water. The majority of these are based on the observation that if the water is aggressive to calcium carbonate, it will also be aggressive to other materials of construction. Furthermore, if conditions are such to encourage the precipitation of calcium carbonate, then this can be used to form a protective calcium carbonate film on the surface of the infrastructure, then this could be considered an effective strategy to mitigate corrosion. As a result most indices use the dissolution or precipitation of calcium carbonate as the basis to determine the stability of the water. This does not however tell the whole picture as water can still be considered corrosive despite having a positive Calcium Carbonate Precipitation Potential (CCPP). The Larson-Skold Index (LI) stands out as the only index that look at other factors such as concentrations of chloride and sulfate ions and their impact on the corrosivity of water to iron and steel pipe, and therefore should always be considered on top of a calcium-carbonate based indices. The most common indices are described below.

3.1 Calcium carbonate precipitation potential (CCPP)

The Calcium Carbonate Dissolution Potential (CCDP) or Calcium Carbonate Precipitation Potential (CCPP) is a most reliable water stability index that is often used in the context of guidelines or regulations without leading to misunderstanding. It provides a quantitative measure of the total amount of calcium carbonate that the water will either dissolve or precipitate giving an accurate guide not only to the nature of the water, but the extent to which it is under saturated with respective to calcium carbonate or over-saturated. The CCPP is an iterative function whose complexity requires the application of computer software for its calculation, but results in the most accurate representation of the water.

When CCPP is calculated, positive values represent a propensity to precipitate calcium carbonate and negative values a propensity to dissolve calcium carbonate. When CCDP is calculated the opposite relationship is formed.

Water is then classified based on the CCPP value (expressed in mg/l $CaCO_3$) as:

- Scale formation for CCPP >10 mg/l
- Protective layer formation for $3 < \text{CCPP} < 10$ mg/l
- Neutral for $-3 < \text{CCPP} < 3$ mg/l
- Mildly corrosive for $-10 < \text{CCPP} < -3$ mg/l
- Aggressively corrosive for $\text{CCPP} < -10$ mg/l

3.2 Langelier saturation index (LSI)

The most commonly used index that provides a measure of the stability of a water with respect to its degree of calcium carbonate saturation is the Langelier Saturation Index (LSI). This is due to the fact that it provides both qualitative representation of the corrosivity of water, and is relatively easy to calculate. First proposed by Prof. WF Langelier in 1936 [14], the Langelier Saturation Index can be calculated as follows:

$$\mathrm{LSI} = \mathrm{pH} - \mathrm{pH_s}$$

where pH_s represents the saturation pH of the water, in which condition the water is in the equilibrium state and neither dissolves, nor precipitates calcium carbonate.

The saturation pH is a complex iterative calculation, similar to the calculation for CCPP, requiring a program to accurately determine it. A simplification of this can be performed by using the ABCD method, which is calculated as follows:

$$\mathrm{pH_s} = (9.3 + \mathrm{A} + \mathrm{B}) - (\mathrm{C} + \mathrm{D})$$

and the parameters defined as:
A = (log [TDS] − 1)/10.
B = −13.12 × log (°C + 273) + 34.55.
C = log [Ca^{+2}].
D = log [Alk].
with TDS expressed in mg/l, Ca^{2+} expressed as mg/l as $CaCO_3$, and Alk expressed as equivalent $CaCO_3$ in mg/l. The plots of **Figure 1** show the slight discrepancy between these two calculation methods.

Waters with positive LSI are oversaturated and tend to form a protective layer of calcium carbonate (scaling effect) on the pipe walls. Highly positive LSI are

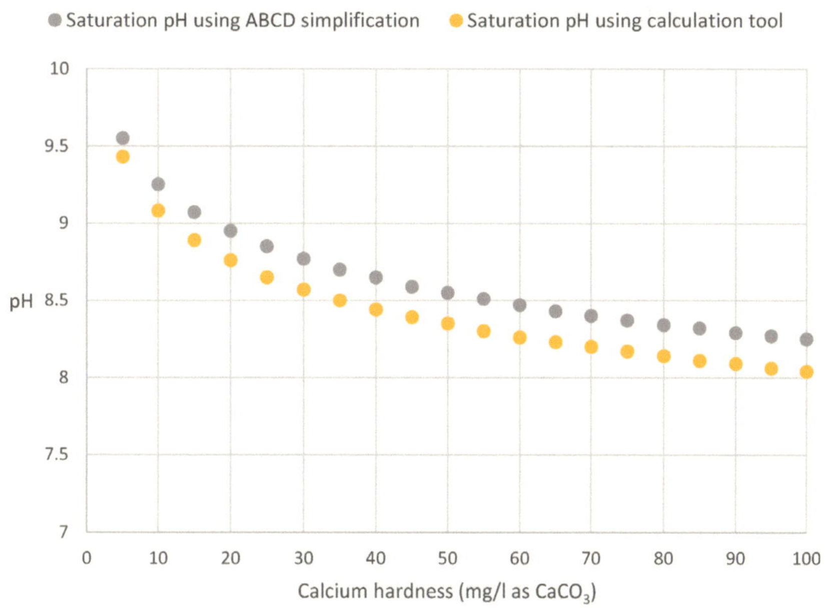

Figure 1.
Relationship between calcium hardness and saturation pH for water with 80 ppm of alkalinity (as $CaCO_3$).

corresponding to high precipitation effect resulting in incrustation. On the other hand, a water with negative LSI value is typically under-saturated with respect to calcium carbonate and so it will potentially dissolve the protective calcium carbonate scale and so be potentially corrosive. LSI alone however, cannot provide an indication of the true indication of the corrosivity of water, as the pH also needs to be considered. A water with LSI of −0.5 at pH 6.0 is much more corrosive than a water with LSI -0.5 at pH 8.0 for example.

LSI is not a reliable indicator of the corrosive tendencies of potable water and that the use of this index together with other models such as empirical determination of chloride, sulfate, alkalinity, dissolved oxygen, buffer capacity, calcium and length of time of exposure would provide information that is more reliable [15].

3.3 Ryznar stability index (RI)

Another parameter similar to the LSI is the Ryznar Stability Index [16], which is looks at the relationship between the saturation pH of the water (with respect to calcium carbonate) and the actual pH of the water. It is given by:

$$RSI = 2pH_s - pH$$

Based on the value assumed by the RSI index, waters are classified as:

- Strongly encrusting, when RSI ranges between 4.0 and 5.0
- Slightly encrusting, when RSI ranges between 5.0 and 6.0
- Slightly corrosive, when RSI ranges between 6.0 and 7.0
- Significantly corrosive, when RSI ranges between 7.0 and 7.5
- Strongly corrosive, when RSI ranges between 7.5 and 9.0
- Unbearably corrosive, when RSI ≥ 9.0

This index provides a reasonably good estimate of expected scale formation even in the presence of phosphate-based inhibitors.

3.4 Puckorius scaling index (PSI)

Similar to both the Langelier Saturation Index and the Ryznar Stability Index, also the Puckorius Scaling Index (PSI) also considers the corrosion potential of the water based on the calcium carbonate saturation of the water [17]. Instead of considering the actual pH of the water however, it defines a new parameter – the equilibrium pH and is defined as:

$$PSI = 2\left(pH_{EQ}\right) - pH_s$$

where, pH_s is the saturation pH as calculated for the previous indices, and pH_{EQ} is the equilibrium pH as calculated by:

$$pH_{EQ} = 1.465 \times \log\,[Alk] + 4.54$$

Thus the water will be:

- Scaling for PSI < 4.5
- Stable for $4.5 \leq PSI \leq 6.5$,
- Corrosive for PSI > 6.5

3.5 Larson-Skold index (LI)

In contrast to the previously descried indices, the Larson–Skold Index (LI) describes the corrosivity of water towards iron or mild steel due to the presence of chloride and sulfate ions [18]. No consideration is given to calcium concentration or pH. This index looks at the relative ratio of chloride and sulfate ions to alkalinity in the water. The reactive anions on one-hand have a strong acidic effect in the anodic pits generated in the exposed corroding metal. The alkalinity due to combined bicarbonate and carbonate ions counter this effect by creating a stabilized and buffered environment that reduces the acidic tendency of the water. The Larson-Skold Index is calculated as follows:

$$LI = \frac{[Cl^-] + [SO_4^{2-}]}{[HCO_3^-] + [CO_3^{2-}]}$$

where the concentrations of each of the species involved is expressed in milliequivalents per liter (meq/L).

The water is then classifies as:

- Non-corrosive for LI < 0.8
- Corrosive for $0.8 \leq LI \leq 1.2$
- Highly corrosive for LI > 1.2

4. Processes used for remineralization of desalinated water

The most common techniques that are currently employed worldwide for water remineralization and stabilization of desalinated or naturally soft water can be divided into three categories: (1) direct dosing of two or more chemical solutions, (2) lime dosing systems, and (3) calcite contactors. Each technique comes with its own benefits and disadvantages, from both a water quality perspective as well as process considerations. Of these three processes, lime dosing systems and calcite contactors see the most widespread application, especially for large desalination plants. This is due to the major drawbacks and costs associated with chemical dosing, as demonstrated below.

4.1 Direct dosing of two or more chemical solutions

One of the simplest and most effective ways of controlling the quality of the final water leaving a treatment facility is through the direct dosing of chemical solutions, either prepared offsite, or involving a simple slurry make down on site when supplied as a solid material. Whilst any combination of chemicals is possible, the most common combination is calcium chloride ($CaCl_2$) for hardness addition in conjunction with sodium bi-carbonate ($NaHCO_3$) to increase the alkalinity of the final water. Expensive sodium hydroxide may in some cases also be needed to reach the required pH. Direct dosing of two or more chemicals has the advantage over other techniques of being able to precisely regulate the quantities of ions added, by controlling the volumes dosed of known solutions. In addition, full dissociation of these highly pure compounds in water avoids any complications of the generation of waste by-products or residual turbidity resulting from low solubility of products.

Despite being a very simple process the greatest drawback of this process is the cost of the chemicals. For this reason, this approach sees very limited application, used at best on small plants where relatively speaking CAPEX has a much greater impact than OPEX.

As an example, in order to achieve 80 mg/l of hardness and alkalinity within the final water (measured as $CaCO_3$), then the required dosage of $CaCl_2$ and $NaHCO_3$ is equivalent to:

$$CaCl_2\left(\frac{mg}{l}\right) = 80\left(\frac{mg}{l}\right) \div 100.08\left(\frac{g}{mol}\right) \times 110.98\left(\frac{g}{mol}\right) = 89\left(\frac{mg}{l}\right)$$

$$NaHCO_3\left(\frac{mg}{l}\right) = 80\left(\frac{mg}{l}\right) \div 100.08\left(\frac{g}{mol}\right) \times 84.006\left(\frac{g}{mol}\right) \times 2 = 134\left(\frac{mg}{l}\right)$$

As two molecular equivalents of $NaHCO_3$ are required to generate one molecular equivalent of alkalinity (measured as $CaCO_3$).

Using the modest values of chemical costs of 300 €/tonne for calcium chloride and 400 €/tonne for sodium bicarbonate (in some locations sodium bicarbonate can be as expensive as 900–950 US$/tonne), then the cost of consumables of this process alone exceeds the total treatment cost of the other treatment processes when energy costs and amortization of investment costs are taken into account (refer **Table 1** below).

Another major drawback of this process and something that is often overlooked, is the increase of the undesired chloride ions that are introduced to the process, as for every mol of calcium that is added, two mol of chloride is also added. This increases the tendency of the water to be corrosive to iron and steel pipe and equipment, as measured and indicated by the previously mentioned Larson-Skold Index.

4.2 Lime dosing systems

Lime dosing systems involve the on-site preparation of a saturated lime solution from either a hydrated lime powder (calcium hydroxide) or quicklime (calcium oxide) which is slaked on-site to produce hydrated lime. The saturated solution is produced by feeding a calcium hydroxide slurry into a lime saturator along with make-up water and a flocculant. The saturator allows for the separation of a clear

	Unit price (EUR/dmt)	Dosage (mg/l)	Cost (EUR/m^3)
Calcium chloride	300	89	0.027
Sodium carbonate	400	134	0.054
TOTAL COST			**0.081**

Table 1.
Chemical costs of calcium chloride and sodium bicarbonate dosing to achieve 80 mg/l of hardness and alkalinity.

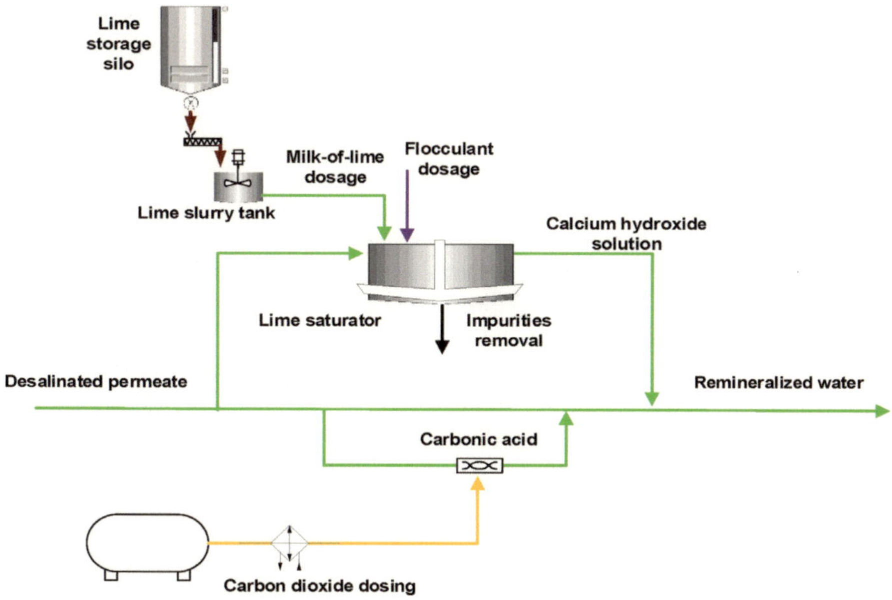

Figure 2.
Schematic of a typical lime dosing system.

solution of calcium hydroxide from the insoluble contents which are settled to the bottom with the aid of the flocculant (**Figure 2**).

The saturated lime solution is then dosed into the final water stream along with carbon dioxide to form bicarbonate alkalinity in the following reaction:

$$Ca(OH)_{2\ (aq)} + 2\ CO_{2\ (g)} \rightarrow Ca(HCO_3)_{2\ (aq)} \tag{2}$$

One of the advantages of lime and reasons that it has been a popular choice for system designers is its worldwide availability as a commercial product and relatively low cost in comparison to other chemicals. It's solubility up to 1700 mg/l at 20°C makes ideal for its preparation within a side stream lending to its smaller footprint relative to calcite contactors.

Lime dosing systems do however have a number of drawbacks and are often the bane of many operators assigned the task to clean and maintain the lime slurry pipework or lime saturators. In comparison to calcium carbonate, lime is more expensive per kilogram of available $CaCO_3$. This can be attributed to the fact that lime is produced by the calcination (burning) and further slaking (hydration) of calcium carbonate which is then dried to produce a powdered hydrated lime. As a result, lime is not only more expensive to produce, but has a much higher carbon footprint. For every kilogram of quicklime that is produced, approximately 700 kcal is required for dissociation and 0.785 kg of carbon dioxide is released [19]. This difference in carbon footprint is not only a concern for plants which strive for environmentally sustainable solutions, but will no doubt further increase the price of lime production as carbon emission taxes are set to play a bigger role in the future.

The operational costs of lime systems are also increased in comparison to calcium carbonate due to the fact that for the same desired quantity of calcium hardness and alkalinity within the final water, twice as much carbon dioxide is required. As can be seen from the chemical reaction equations, remineralization using calcium hydroxide requires two molecular equivalents of carbon dioxide for each mole of calcium hydroxide (refer Eq. (2)). Remineralization using calcium carbonate however, requires only one molecular equivalent of carbon dioxide for each mole of calcium carbonate (refer Eq. (3)). Carbon dioxide is in most cases the largest operating cost for post treatment systems, so this has an important impact on the overall cost per cubic meter of treated water.

As mentioned previously, hydrated lime contains an insoluble content. This insoluble content is for the most part is unburnt calcium carbonate but can also include silicates and other impurities. These impurities are usually in the order of 5–15% and must be removed and dealt with as a waste product. Needless to say the lower the insoluble content, the purer the product and the higher the price of the product. If the impurities are not effectively removed, they will add to the turbidity in the final water. To improve the efficacy of the clarification process, a flocculant is often dosed to aid in the settling. This waste then needs to be thickened on site and sent away for proper disposal. These factors further add to the operational cost and complexity of lime dosing systems. The operation of a lime clarifier is very sensitive to factors such as temperature, flocculent dosing, and throughput flow rate, meaning they do not handle fluctuations in plant flow very well. Even a perfectly functioning lime clarifier can still be susceptible to turbidity problems. Absorption of carbon dioxide from the atmosphere can lead to the increase in dissolved inorganic carbon within the lime solution resulting in precipitation of calcium carbonate and producing a cloudy solution.

Extending the example for chemical dosing, in order to achieve 80 mg/l of hardness and alkalinity within the final water (measured as $CaCO_3$), then the required dosage of $Ca(OH)_2$ and CO_2 is equivalent to:

$$Ca(OH)_2\left(\frac{mg}{l}\right) = 80\left(\frac{mg}{l}\right) \div 100.08\left(\frac{g}{mol}\right) \times 74.1\left(\frac{g}{mol}\right) \div 90\% = 65\left(\frac{mg}{l}\right)$$

(assuming a $Ca(OH)_2$ purity of 90%)

$$CO_2\left(\frac{mg}{l}\right) = 80\left(\frac{mg}{l}\right) \div 100.08\left(\frac{g}{mol}\right) \times 44.01\left(\frac{g}{mol}\right) \times 2 = 70\left(\frac{mg}{l}\right)$$

As two molecular equivalents of CO_2 are required to generate one molecular equivalent of alkalinity (measured as $CaCO_3$). Considering also 10% of product is removed as waste from the lime saturator, and then dewatered to a maximum of 30% solids:

$$Waste\ generated\ \left(\frac{mg}{l}\right) = 65\left(\frac{mg}{l}\right) \times 10\% \div 30\% = 22\left(\frac{mg}{l}\right)$$

This generates treatment costs that are a fraction of that required for chemical dosing as presented below in **Table 2** below.

4.3 Calcite contactors

Remineralization of demineralized water using calcite contactors is achieved by passing a stream of acidified water through a bed of calcite chips. These chips are

	Unit price (EUR/dmt)	Dosage (mg/l)	Cost (EUR/m^3)
Calcium hydroxide	250	65	0.016
Carbon dioxide	200	70	0.014
Waste disposal	50	22	0.001
TOTAL COST			**0.031**

Table 2.
Approximate operational costs for lime dosing systems to achieve 80 mg/l of hardness and alkalinity.

usually limestone or marble, but in some cases dolomite, chalk, precipitated calcium carbonate or even muscle shells is used. The acidified water dissolves the calcium carbonate, increasing the calcium hardness of the water, the carbonate alkalinity and the pH through the following reaction (when carbon dioxide is used as the acidifying agent) (**Figure 3**):

$$CaCO_{3\ (s)} + CO_{2\ (g)} + H_2O \rightarrow Ca(HCO_3)_{2\ (aq)} \qquad (3)$$

As noted earlier, calcium carbonate is one of the most common minerals available, taking up almost 4% of the earth's crust [10]. As a result it is a readily sourced product and processing requirements are minimal as the chemical composition does not need to be altered before use. Consequently, and as alluded to previously, calcium carbonate can be supplied at a lower cost than lime. In addition, it requires half as much carbon dioxide for the same quantity of calcium bi-carbonate (refer Eqs. (2) and (3)). Added to this, calcium carbonate used for water remineralization can be very pure, with an insoluble content as little as 0.1%. This further reduces operating costs, with more available product for use and less produced waste which requires further handling and disposal. Calcium carbonate is also chemically stable and non-corrosive making it easy for manual handling. In comparison, calcium hydroxide is classified as hazardous, and exposure can cause burning and irritation.

The main disadvantage of calcium carbonate is its chemical solubility which is only 13 mg/l in pure water, and requires the addition of an acid to dissolve quantities above this. As the water gets closer to the saturation point, the rate of reaction slows down dramatically. As a result, thermodynamic equilibrium is almost impossible to reach in such a dissolution reactor [20], and requires excessive contact time between the water and the calcite bed. This increases the overall size of the plant required to treat a certain volume of water and the resultant capital investment.

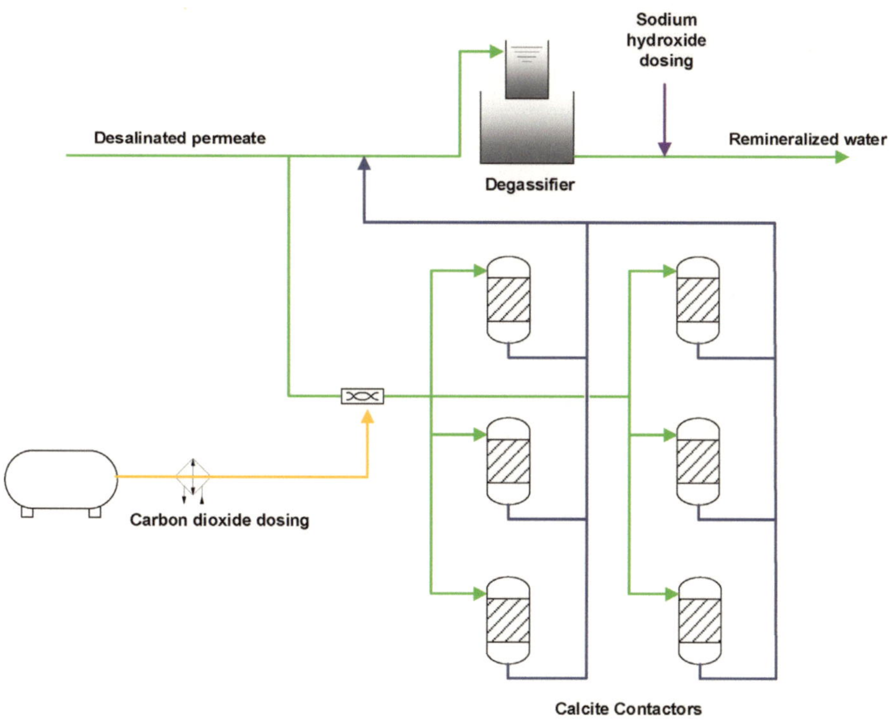

Figure 3.
Schematic of typical calcite contactor process.

In order to increase the rate of reaction and circumvent this problem, the pH of the feed water is often decreased before the calcite contactor, in excess of what would normally be required. This results in a faster dissolution rate so that the required hardness and alkalinity increase occurs in a shorter time period. The water however, does not achieve saturation levels with respect to calcium carbonate, as the pH of the effluent leaving the reactor is much lower than the saturation pH for its calcium carbonate content. This also decreases the efficiency of the process described by Eq. (3), as it requires excess carbon dioxide to be dosed for the increased acidity. This not only increases the chemical costs for the process, but produces effluent leaving the reactor with a quantity of unreacted CO_2. The pH of the water must then be adjusted either by the dosage of a strong base (e.g. sodium hydroxide) or the liberation of carbon dioxide to achieve a zero or slight positive LSI (Langelier Saturation Value) value, as required by most treatment facilities.

This pH adjustment step adds to both the investment costs: due to the requirement of additional infrastructure for stripping equipment or an additional chemical dosing system, and the operational costs: due either to additional chemical consumption when a strong base is used, or additional power consumption when the excess CO_2 is stripped. In some instances both are required. Sodium hydroxide is a relatively expensive chemical, and even a small dosage can add significantly to the overall cost of the remineralization process. In many cases, it is most cost effective to waste the excess CO_2 through stripping rather than convert it to additional alkalinity through the dosage of a strong base. For plants that require a low level of remineralization though, (e.g. 50 mg/l) the quantity of free of CO_2 that remains in the water at saturation level is so low that this cannot be achieved through stripping alone and requires some dosage of a strong base for partial or total pH adjustment.

Other issues facing operators of calcite contactors are turbidity spikes in the treated water leaving the contactor. These are generally a result of the introduction of new material to the calcite reactor and the accompanying "fines" supplied with the raw material. To deal with this calcite contactors require frequent backwashes, in particular following the loading of new material. This adds an additional need for a backwash treatment and handling system at plant to deal with this waste stream. The change in bed heights between fills also result in a change in water quality leaving the calcite contactor due to variances in the quantity of product available for reaction. In most cases calcite contactors are loaded manually, increasing the operational costs due to additional operator utilization. Lime systems on the other hand are fed automatically from a lime storage silo, requiring operator input only to receive new deliveries. Calcite contactors also require regular backwash with both air and water. This serves to remove excess fines and insoluble waste material caught on the calcite chips as well as resettling the calcite bed to prevent preferential flow paths of the water through the bed.

Completing the cost comparison example to achieve 80 mg/l of hardness and alkalinity within the final water (measured as $CaCO_3$), then the required dosage of $CaCO_3$ and CO_2 is equivalent to:

$$CaCO_3\left(\frac{mg}{l}\right) = 80\left(\frac{mg}{l}\right) \div 99\% = 65\left(\frac{mg}{l}\right)$$

Assuming a $CaCO_3$ purity of 99%,

$$CO_2\left(\frac{mg}{l}\right) = 80\left(\frac{mg}{l}\right) \div 100.08\left(\frac{g}{mol}\right) \times 44.01\left(\frac{g}{mol}\right) \times 130\% = 46\left(\frac{mg}{l}\right)$$

Assuming 30% extra carbon dioxide is required to increase the rate of reaction, and finally an addition of 2.5 ppm of sodium hydroxide is required after degassing to bring the pH to saturation conditions.

	Unit price (EUR/dmt)	Dosage (mg/l)	Cost (EUR/m^3)
Calcite chips	100	81	0.008
Carbon dioxide	200	46	0.009
Sodium hydroxide	800	2.5	0.002
TOTAL COST			**0.019**

Table 3.
Approximate operational costs for calcite contactors to achieve 80 mg/l of hardness and alkalinity.

The approximate treatment costs for calcite contactors are shown below in **Table 3** demonstrating that they are not only less expensive to operate than lime dosing systems, but they offer a more environmentally friendly solution. The big drawback for calcite contactors, which is not demonstrated here are the large physical footprint and investment required. These elements alone can often drive a designers decision towards lime dosing system, particularly in locations such as Singapore, where space is a premium.

5. Latest developments and trends

Historically, the main focus for developments within the field of desalination has been the improvement and optimization of the reverse osmosis and pre-treatment processes. This is due to the fact that these areas comprise the largest fractions of capital investment and are responsible for the largest portion of operating costs. In comparison, only minor inroads have been made to improve and optimize post treatment processes. In spite of this, the value and importance of post treatment processes should not be underestimated, as it is these processes that are ultimately responsible for the final water quality sent to the consumer.

Whilst calcite contactors have many advantages over lime dosing systems, their major drawbacks are centered around their slow reactivity which result in large physical footprint and investment for the dissolution reactors. In order to address the issue of slow dissolution kinetics of calcite chips, new and innovative processes have been developed over the last few years that utilize micronized calcium carbonate. These processes take advantage of the increased surface area and reaction kinetics available from the micronized products to achieve decreased contact times, higher concentrations, improved carbon dioxide efficiency, or a combination of these factors. Micronized calcium carbonate is dissolved in a Membrane Calcite Reactor (MCR) which combines a submerged ultrafiltration membrane immersed in a suspension of micronized calcium carbonate. Carbon dioxide is added to the calcium carbonate suspension, which in turns reacts to form a calcium bi-carbonate solution. The membrane acts as a barrier between the dissolved and undissolved calcium carbonate enabling a perfectly clear solution to be extracted from the reactor that can be dosed into the desalination permeate. The use of micronized calcium carbonate results in fast reaction times, and hence decrease footprint and investment offering an improvement over current processes (**Figure 4**).

Although remineralization processes primarily concern themselves with the replenishment of calcium hardness and alkalinity, more recently attention has been given to the need to replenish magnesium ions, with some countries considering the implementation of legislation for these purposes. Magnesium is arguably the most important mineral for the body, being utilized by every organ, in particular the heart, muscles and kidney. It is the fourth most abundant cation in the body, and the second most in intracellular fluid [21]. Magnesium deficiency has also been

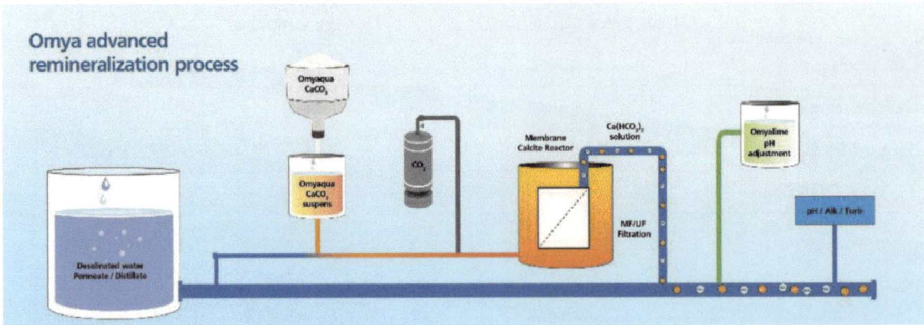

Figure 4.
Schematic of the Omya advanced remineralization process (OARP) based on micronized calcium carbonate.

scientifically proven to either trigger or cause the following health problems: heart disease, diabetes, migraines, anxiety, hypertension, depression, fatigue, blood clots, liver disease, kidney disease, osteoporosis, insomnia, fatigue, cystitis, nerve problems and hypoglycemia [22]. Despite these facts and the relative importance of magnesium, up to 75 percent of people do not receive the recommended daily intake of magnesium (based on studies performed in the US – global intakes vary greatly based on local diets) [23].

Magnesium deficiency has also been specifically linked to higher rates of mortality in terms of cardiovascular deaths as well as general mortality. A recent German study sampling over 4000 people showed a strong correlation between low serum magnesium levels (< 0.73 mmol/l) and cardio-vascular deaths at a rate of 3.44 deaths per 1000 person years, in comparison to 1.53 deaths per 1000 person years for those with higher magnesium concentrations. More importantly though, an even stronger correlation was found between serum magnesium levels and all-cause deaths. For those with low serum magnesium levels, the mortality rate was 10.95 deaths per 1000 person years compared to 1.45 deaths per 1000 person years at higher serum magnesium concentrations [24]. Researchers from the Bar Ilan University together with the Tel HaShomer Hospital gave more weight to this argument based on their review of death rates in Israel in areas serviced by desalinated water in comparison to those supplied by natural water. In their study they noted a marked difference in the number of deaths from heart disease in the areas that were supplied water from desalination in comparison to those supplied by natural water which had not been previously recognized when comparisons were made before desalination was introduced [25]. The conclusion was drawn by the researcher that this was as a result of decreased magnesium intake in these areas after the introduction of desalination. Only causal links however were established with direct links still to be proven.

Furthermore when reviewing the total number of epidemiological studies from the 1950's until present that had been performed on the link between magnesium in drinking water and cardiovascular mortality, it was determined that there is enough evidence to support a link, especially for concentrations above 5 mg/l [26]. The World Health Organization (WHO) also recommends maintaining a minimum Mg^{2+} concentration of 10 mg/l in all drinking waters [27]. Despite these recommendations, the replenishment of magnesium salts is rarely performed, if at all. One of the main obstacles is the relative cost of current methods, which significantly increases the total cost to desalinate and stabilize the water. The addition of magnesium to drinking water is commonly achieved through the dosing of chemical solutions such as magnesium chloride or magnesium sulphate. The high solubility of both salts allows for the supply of highly concentrated solutions, or simple solution make-down systems on site using crystalline salts, and the accurate dosing of these solutions to produce the magnesium concentration in the final water as desired. Whilst this process is very effective, it is also extremely expensive. This tends to position this subject as a question of luxury rather than necessity. The use of these chemicals in fact renders an additional cost to treat the drinking water by as much as 10 US cents/m^3 [28]. For this reason, more cost-effective replenishment methods are becoming of high interest in order to be prepared to face the coming soon changes in drinking water regulation.

The use of natural minerals for the replenishment of magnesium offers both a low cost and sustainable alternative to chemical dosing. Like natural calcium carbonate, magnesium minerals are found within the earth's crust as a range of sparingly soluble compounds, that naturally replenish themselves through dissolution and precipitation cycles. The fact that these minerals are sparingly soluble, increases their prevalence and concentrations in nature, as they are more likely to precipitate from solution than their highly soluble counterparts such as magnesium chloride and magnesium sulphate, which are only found in limited locations such as the dead sea, and in these cases require further refining. The limited solubility of the minerals means that they alone, struggle to provide the required levels of dissolved magnesium without the addition of an acid, to increase both their rate of dissolution and total concentration. Additionally, because the low solubility and slow reaction kinetics of these minerals, large contact tanks and an expensive installation are often required to achieve the required levels of dissolution. These issues can be effectively countered through the dissolution of powdered products within Membrane Calcite Reactors, similar to that for calcium carbonate.

Alternatively, some companies and research institutions are investigating the "mining" of natural resources from the brines rejected by desalination processes. These are often rich in a number of metals and minerals that are essential for industry and otherwise not scarce in availability [29]. These include magnesium, scandium, vanadium, gallium, boron lithium, indium, molybdenum and rubidium. These processes could offer a cheap source of magnesium at desalination sites where they could be immediately reinjected into the final water to replenish some of what has been extracted.

Author details

Nicholas Nelson* and Antonella De Luca
Omya International, Oftringen, Switzerland

*Address all correspondence to: nicholas.nelson@omya.com

References

[1] McNeill L, Edwards M. Iron Pipe Corrosion in Distribution Systems. AWWA Journal. 2002; 93/7:88-100. DOI: 10.1002/j.1551-8833.2001.tb09246

[2] Schraper D. As Infrastructure Crumbles, Trillions of Gallons of Water Lost, NPR News; 29 October 2014. Available from: https://www.wnpr.org/post/infrastructure-crumbles-trillions-gallons-water-lost [Accessed: 2021-06-03]

[3] Lahav O, Birnhack L. Quality Criteria for Desalinated Water Following Post-Treatment. Desalination. 2007;207: 286-303. DOI: 10.1016/j.desal.2006.05.022

[4] US Environmental Protection Agency. Lead in Drinking Water: Basic Information. 2004. Available from: http://www.epa.gov/safewater/lead/basicinformation.html [Accessed: 2021-06-03]

[5] World Health Organization. Cadmium in Drinking-water: Background Document for Development of WHO Guidelines for Drinking-water Quality. 2011. Available from: https://www.who.int/water_sanitation_health/dwq/chemicals/cadmium.pdf [Accessed: 2021-06-03]

[6] Guo Q, Toomuluri PJ, Eckert Jnr. JO. Leachability of Regulated Metals from Cement-Mortar Linings. AWWA Journal. 1998;90,7:62-73. DOI: 10.1002/j.1551-8833.1998.tb08397

[7] World Health Organization. Guidelines for Drinking Water Quality. 4th ed, 2011. Available from: http://who.int./water_sanitation_health/publications/dwq-guidelines-4/en/ [Accessed: 2021-06-03]

[8] Sontheimer H, Kolle W, Snoeyink VL. Siderite model of the formation of corrosion-resistant scales. AWWA Journal. 1981;73:572. DOI: 10.1002/j.1551-8833.1981.tb04801

[9] Oliphant RJ, Causes of Copper Corrosion in Plumbing Systems, Foundation for Water Research. 2017. Available from: http://www.fwr.org [Accessed: 2021-06-03]

[10] Tegethoff WF, Rohleder J, Kroker E. Calcium Carbonate. Birkhäuser Verlag; 2001. DOI: 10.1007/978-3-0348-8245-3

[11] World Health Organization. Hardness in Drinking-water: Background Document for Development of WHO Guidelines for Drinking-water Quality. 2011. Available from: https://www.who.int/water_sanitation_health/dwq/chemicals/hardness.pdf [Accessed: 2021-06-03]

[12] Łagocka R, Sikorska-Bochińska J, Noceń I, Jakubowska K, Góra M, Buczkowska-Radlińska J. Influence of the Mineral Composition of Drinking Water Taken from Surface Water Intake in Enhancing Regeneration Processes in Mineralized Human Teeth Tissue. Pol. J. Environ. Stud. 2011;20(2):411–416.

[13] World Health Organization. Guidelines for Drinking Water Quality. 4th ed, 2011. Available from: http://who.int./water_sanitation_health/publications/dwq-guidelines-4/en/ [Accessed: 2021-06-03]

[14] Langelier WF. Chemical Equilibria in Water Treatment. AWWA Journal. 1946; 38:169–178. DOI: 10.1002/j.1551-8833.1946.tb17557

[15] Pisigan Jr. RA, Singley JE. Evaluation of Water Corrosivity Using the Langelier Index and Relative Corrosion Rate Models. Corrosion Journal, National Association of Corrosion Engineers. 1984;149. DOI: 10.1002/j.1551-8833.1985.tb05647

[16] Ryznard JW. A new index for determining amount of calcium carbonate scale formed by a water. AWWA Journal. 1944;36:472–486. DOI: 10.1002/j.1551-8833.1944.tb20016

[17] Puckorius R, Brooke JM. A new practical index for calcium carbonate scale prediction in cooling systems. Corrosion Journal. 1991;47:280–284. DOI: 10.5006/1.3585256

[18] Larson TE, Skold RV. Laboratory Studies Relating Mineral Quality of Water to Corrosion of Steel and Cast Iron. Corrosion Journal. 1958;71:43–46. DOI: 10.5006/0010-9312-14.6.43.

[19] Kumar GS, Ramakrishnan A, Hung YT. Lime Calcination, Advanced Physicochemical Treatment Technologies. Humana Press. 2012: 611-633. https://doi.org/10.1007/978-1-59745-173-4_14

[20] Lahav O, Voutchkov N, Birnhack, L. Post-Treatment of Desalinated Water, Balaban Desalination Publications. 2012

[21] World Health Organization. Hardness in Drinking-water: Background Document for Development of WHO Guidelines for Drinking-water Quality. 2011. Available from: https://www.who.int/water_sanitation_health/dwq/chemicals/hardness.pdf [Accessed: 2021-06-03]

[22] Dean C. The Magnesium Miracle. Ballantine Books. 2014

[23] Barnes Z. Magnesium, an invisible deficiency that could be harming your health. Life by DailyBurn - CNN, 3. January 2016. Available from: https://edition.cnn.com/2014/12/31/health/magnesium-deficiency-health/index.html. [Accessed: 2021-06-03]

[24] Reffelmann T, Ittermann T, Dörr M, Völzke H, Reinthaler M, Petersmann A, Felix SB. Low serum magnesium concentrations predict cardiovascular and all-cause mortality. Atherosclerosis 2011;219.1:280-284. DOI: 10.1016/j.atherosclerosis.2011.05.038

[25] Findel H. What To Do? Desalination Found to Cause Fatal Heart Disease. Israel National News. 7. April 2016. Available from: https://www.israelnationalnews.com/News/News.aspx/210501 [Accessed: 2021-06-03]

[26] Calderon R, Hunter P. Epidemiological studies and the association of cardiovascular disease risks with water hardness in Calcium and Magnesium in Drinking-Water: Public Health Significance, World Health Organization, 2009; p108-142. Available from: https://www.who.int/water_sanitation_health/publications/publication_9789241563550/en/ [Accessed: 2021-06-03]

[27] World Health Organization. Guidelines for Drinking Water Quality. 4th ed, 2011. Available from: http://who.int./water_sanitation_health/publications/dwq-guidelines-4/en/ [Accessed: 2021-06-03]

[28] De Luca A, Bouton A, Nelson N, Kallenberg J. The Use Of Natural Minerals For The Replenishment Of Magnesium Into Drinking Water Following Desalination, In: The IDA World Congress on Desalination and Water Reuse; 20-24 October 2019; Dubai, UAE

[29] Sea4Value – Mining Value from Brines [Internet]. 2021. Available from: http://sea4value.eu [Accessed: 2021-06-03]

Chapter 6

Elimination of Acid Red 88 by Waste Product from the Phosphate Industry: Batch Design and Regeneration

Khaled Boughzala and Mustapha Hidouri

Abstract

Waste regenerated after washing of rock phosphate and phosphogypsum has been proposed as removal agents of Acid Red 88 (AR 88)) from artificially contaminated solution. Natural phosphate (PN) was also studied for comparison. These materials were characterized beforehand, as is intended for the removal tests, by chemical analysis, powder X-ray diffraction, Fourier-transform infrared spectroscopy, thermogravimetric analysis– differential thermal analysis, scanning electron microscopy, and N_2 adsorption isotherms. The conducted experiments show that among the different materials, the PWR has the highest retention capacity of the dye (123.4 mg g^{-1}) of AR-88. Upon calcinations, the removal capacities reduced by 60 to 70%. We take note also that a decrease in the amount of removed AR 88 dye occurs with an increase in pH. The kinetics data on the reaction between AR 88 and the materials are described well by a pseudo -second-order model. The Langmuir model is successfully applied to the experimental data of the removal of acid red 88. The removal process is exothermic.

Keywords: Natural phosphate, Phosphogypsum removal, Acidic Red 88

1. Introduction

Synthetic dyes have been used in several industrial sectors such as the automotive sector, the textile industries, leather tanning, plastics, aper, and photoelectrochemical cells and therefore, A significant amount of water is used [1]. Wastewater laden with dyes is usually dumped in sewers, rivers, and nearby lagoons. Such treatment affects the water quality, the aquatic ecosystem and the biodiversity of the environment [2, 3]. Wastewater treatment is necessary before it is released into the environment [4]. Likewise, wastewater treatment has been proposed as a solution to obtain good quality water for agricultural and industrial applications [5, 6]. The literature reports several techniques for the treatment and depollution of textiles effluents. It is important to mention, among them, the membrane filtration techniques [7], coagulation / flocculation [8], electro- coagulation [9], oxidation techniques [10] and aerobic biological processes and anaerobes [11, 12]. The search for other effective methods is necessary because these previously described processes have financial limitations and design complexities.

The adsorption process is used for water treatment, however, the high price of adsorbent materials, remains a constraint [13, 14].

Among the proposed solutions include using adsorbents from agricultural residues, for example, the waste material of corn cob, palm fruit parts, the chestnut peel, almond shell, rice husk, orange, and lime peels, pine fruit shells, and others [15, 16].

The clay minerals and the waste produced by the phosphate industries have been used for the retention of textiles dyes [16–21]. The phosphate rocks is Tunisia is considered among the top countries that produce phosphate rocks [22]. Tunisia is ranked among the top phosphate-producing countries. The Natural phosphate (NP) is an abundant product extracted from phosphate rocks and made up of a carbonated fluoroapatite with important substitution of phosphate by carbonate [23]. The literature reveals the existence of a few studies related to the adsorption of some dyes in polluted textile water or adsorption of basic dyes and reactive dyes using natural phosphates (NP) [24–26].

The calcination (heat treatment) of natural phosphate rocks was described in the literature for upgrading the calcareous phosphate ores [27], and to propose other applications than direct synthesis of fertilizers, such the production of pure chemicals, soft drinks and pharmaceutical products [28].

Although the properties of the calcined materials were well documented for the synthesis of fertilizers, however, there are no attempts to describe their usage for the elimination of dyes from polluted water.

Also, the phosphate rocks are converted into phosphoric acid by the addition of sulfuric acid through the socalled wet process [29]. The production of phosphoric acid resulted in the formation of huge amount of wastes by products such as phospohogypsum [30]. Some evaluation of theses wastes were proposed in the treatment of polluted water and some research activities have studied their utility as removal agents of some dyes and heavy metal ions from aqueous solutions [31–40].

2. Materials and methods

2.1 Adsorbents

The NP was picked up in the Gafsa-Metlaoui basin (appointed the NP sample). PG is a by-product result of the reaction of sulfuric acid and phosphate rock. PWR is a byproduct of a phosphate company's washing plant. The samples were washed and then heated in an oven at 105° C. The acid dye AR 88 was purchased from ATUL Limited, India; The maximum absorption of this dye is examineted at the wavelength of 508 nm. Co $(NO_3)_2 \times 6H_2O$ and oxone ($2KHSO_5$ $KHSO_4$ K_2SO_4) were used without further treatment after purchase from Alfa Aesar (4.7% active oxygen) (Alfa Aesar, Lancashire, UK United).

2.2 Adsorption experiments

Different solutions with concentrations between 5 and 200 mg L^{-1} was prepared by diluting a solution of an initial concentration of 1000 mg L^{-1}. Then, a mass of 0.1 g of the adsorbent was introduced to get to a total volume of 200 mL (dye solution) with a shacking speed equal to 150 rpm, natural pH, and room temperature. The retained quantity (qe, mg g–1) and the removal percentage (% R) were calculated by Eqs. (1) and (2), respectively.

$$q_t = (C_0 - C_t)\frac{V}{m} \tag{1}$$

$$P(\%) = 100 * \frac{(C_0 - C_t)}{C_0} \tag{2}$$

where Ci and Ce correspond to the initial and equilibrium concentrations of the anionic dye (mg L^{-1}). V is the employed solution volume (L) and m is the adsorbent mass (mg). For to examine the effect of various parameters on the retention of the AR 88 dye, the dosage of adsorbents, kinetic, initial dye concentrations, pH, and temperature were studied independently.

2.3 Regeneration of used materials

NP, PG, or PWR samples were mixed to 200 mL of a fresh solution of AR 88 (Ci = 200 mg L^{-1}). The mixture is left for 24 hours. The spent adsorbent was obtained by centrifugation and treated with a 10 mL solution of $Co(NO_3)_2 \times 6H_2O$ and 12 mg of oxone ($2KHSO_5 \times KHSO_4 \times K2SO_4$). The regenerated adsorbent was centrifuged, washed a few times with deionized water, and then reused in the next run. The solution of Co and oxone was not discharged and used in the next recycle runs [26].

2.4 Characterization

The chemical composition of the used products was determined using atomic absorption spectroscopy (Perkin-Elmer 3110, Waltham, Massachusetts USA).

X-ray diffraction (XRD) patterns were carried on a X'Pert Pro, PANalytical diffractometer (Malvern, United Kingdom) with Cu K radiation.

The data were collected in a 2θ range from 5° to 80°, with a step size of 0.02° and a scanning step time of 10 s. The mineral phases were identified from the data given in the American Society for Testing and Materials cards. FTIR spectra were recorded with a Perkin Elmer 1283 spectrometer (Waltham, Massachusetts USA) in the range of 3500–350 cm^{-1} using samples pressed into pellets with KBr.

Thermal analysis was conducted in air from room temperature to 1000°C at a heating rate of 10°/min, using a a Setaram Instrumentation (Caluire - France) SETSYS. The surface morphology of the materials was observed by scanning electron microscopy (SEM, FEI Quanta 200, Hillsboro, Oregon, USA). The specific surface area values were estimated from nitrogen adsorption isotherms using the Brunauer–Emmett– Teller (BET) equation. The isotherms were determined using a Micromeritics ASAP 2020 system (Norcross, Georgia, USA). The compounds were outgassed at 120°C for 8 h prior to the measurement. The pHzpc of the NP, PG, and PWR samples was measured in solutions of NaCl (0.01 mol L^{-1}). The concentration of AR 88 at equilibrium was mesured during the removal by a UV–visible spectrophotometer (Perkin- Elmer model LAMBDA20, Waltham, Massachusetts USA) at maximum wavelength of 508 nm.

3. Results and discussion

3.1 Characterization of used materials

The values of the chemical analysis are presented in **Table 1**. It is It is remarked that the high CaO concentration (45%) is observed in natural phosphate (NP) compound and it diminished at 26.7% in the phosphate waste rock compound

Samples	$\%P_2O_5$ (%)	CaO (%)	MgO (%)	Cd (ppm)	CaO/P_2O_5
NP	25.64	44.94	0.87	45	1.75
PG	4.06	36.68	0.53	15	0.34
PWR	14.01	26.72	2.15	51	1.90

Table 1.
The chemical analysis of the main elements in the three samples.

(PWR). The NP sample has a the maximum value of P_2O_5 content (25.6%) compared to PWR compound (14.0%).

The concentration of anhydride phosphorus pentoxide in NP was close to that reported for similar rocks [33, 34]. The diminish of CaO and P2O5 contents was related to washing process. A small percentage of MgO, from 0.87% to 2.15%, was observed. The Ca/P atomic ratio is about 1.75 for NP sample, and close 1.91 for PWR [41, 42]. This variation was due to lower amount of P2O5 in PWR, resulted from the washing process. The Cd quantity is located between 45 and 51 ppm [43].

Figure 1 shows the powder XRD patterns of the phosphate waste rock, phosphogypsum, and phosphate waste rock samples. The natural phosphate and phosphate waste rock patterns exhibit similar patterns. Mineralogical identification reveals the presence of carbonate fluorapatite $Ca_{9.55}(PO_4)_{4.96}F_{1.96}(CO_3)_{1.28}$ and other materials, such as heulandite $((C_2H_5)NH_3)_{7.85}(Al_{8.7}Si_{27.3})O_{72})$ $(H_2O)_{6.92})$ and quartz (SiO_2) [34]. The sample of phosphogypsum exhibits different phases, such as bassanite ($CaSO_4{\cdot}½H_2O$) and anhydrite ($CaSO_4$) compounds.

The IR absorption spectra of the different samples are displayed in **Figure 2**. The samples present the absorption bands associated to the PO_4^{3-}groups between 1042, 570, 520 and 470 cm^{-1}. These frequencies correspond to the vibration modes ν_3, ν_4, and ν_2, respectively [44].

The FTIR spectrum of the phosphogypsum is shown in **Figure 2**. The spectrum is identified by the typical absorption bands reported for the gypsums compounds [45]. These bands were observed at 1120 cm^{-1} 600–660 cm^{-1} (υ_4), and 470 cm^{-1} (υ_2).

The doublet at 1463 and 1426 cm^{-1}, and the band at about 863 cm^{-1} were assigned to the υ_3 and υ_2 vibration modes of CO_3^{2-} groups [46]. These bands indicate the existence of CO_3 groups in the gypsum structure. The bands characterized ted to SiO_2 products are similar to the bands of PO_4 groups at 1042 cm^{-1}, with

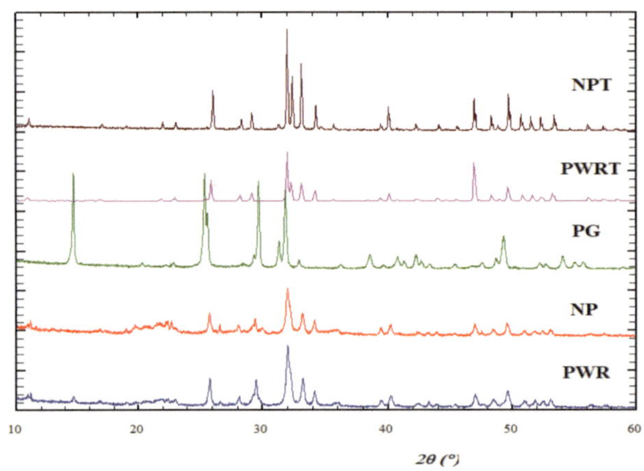

Figure 1.
Powder XRD patterns of (a) natural phosphate, (b) phosphogypsum and (c) phosphate waste rock. (A) corresponds to anhydrite, (B) to bassanite, (H) to heulandite.

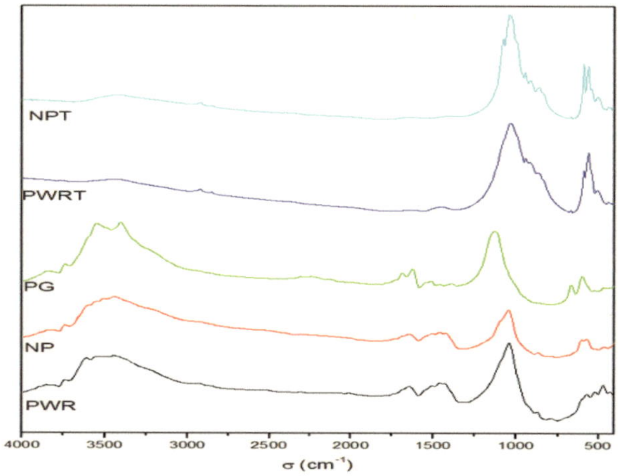

Figure 2.
FTIR spectra of (a) natural phosphate, (b) phosphogypsum, and (c) phosphate waste rock.

a shoulder at 474 cm^{-1}. The bands assigned to adsorbed water molecules at 3600 and 1600 cm^{-1} are present on some spectra.

The thermal gravimetric analysis (TGA) and differential thermal analysis (DTA) curves of the natural phosphate and phosphate waste rock samples are given in **Figure 3**. The TGA curve of the natural phosphate and phosphate waste rock samples exhibits three consecutive mass losses. The first mass loss, observed between room temperature and 150°C, is related to the water desorption.

The second mass loss step related to the loss of water content and dehydroxylation from 170–450°C. The third mass losse starts at 400°C and continues to 1,000°C, is attributed to the decomposition of carbonates and other materials [47]. The two first mass losses are associated with endothermic effects that is observed on the differential thermal analysis (DTA) curve at 85°C, 100°C, 140°C, and 350°C [48]. The Third loss mass is accompanied by two broad exothermal peaks are observed in the range of 700–750°C and are attributed to the phase transformation of some resulting samples.

TG/DTA curves of the phosphogypsum product are given in **Figure 3**. The first weight loss observed between 120°C and 350°C is due to the elimination of the entire water of crystallization. The second weight loss achieved between 300°C and 450°C was attributed to the decomposition of $CaSO_4$ to CaO [49].

The elimination of the entire water of crystallization is related to an endothermal peak at 207°C. The DTA curve exhibits also a broad peak with low intensity.

The scanning electron microscope micrographs of the natural phosphate and phosphate waste rock samples show the presence of nonporous particles of different sizes with spherical shapes or ovoid grains. Also, for the phosphogypsum product, several shapes, such as hexagonal, tabular, and needle-like, are examined (**Figure 4**).

The specific surface areas (SBET) of the three compounds samples are 16.39, 11.35, and 26.02 $m^2\ g^{-1}$, respectively, and reveal the nonporous character of these adsorbents.

The slight rise in the SBET value of the phosphate waste rock compound could be interpreted by the acid activation of these rocks.

These values are near to those studied in the case of natural phosphate rocks [50]. The average pore volumes are varied between 0.023–0.053 cc g^{-1}. The average pore diameter is in the interval of [9.58–7.62] nm, which proves the nonporous character of the studied samples (**Table 2**).

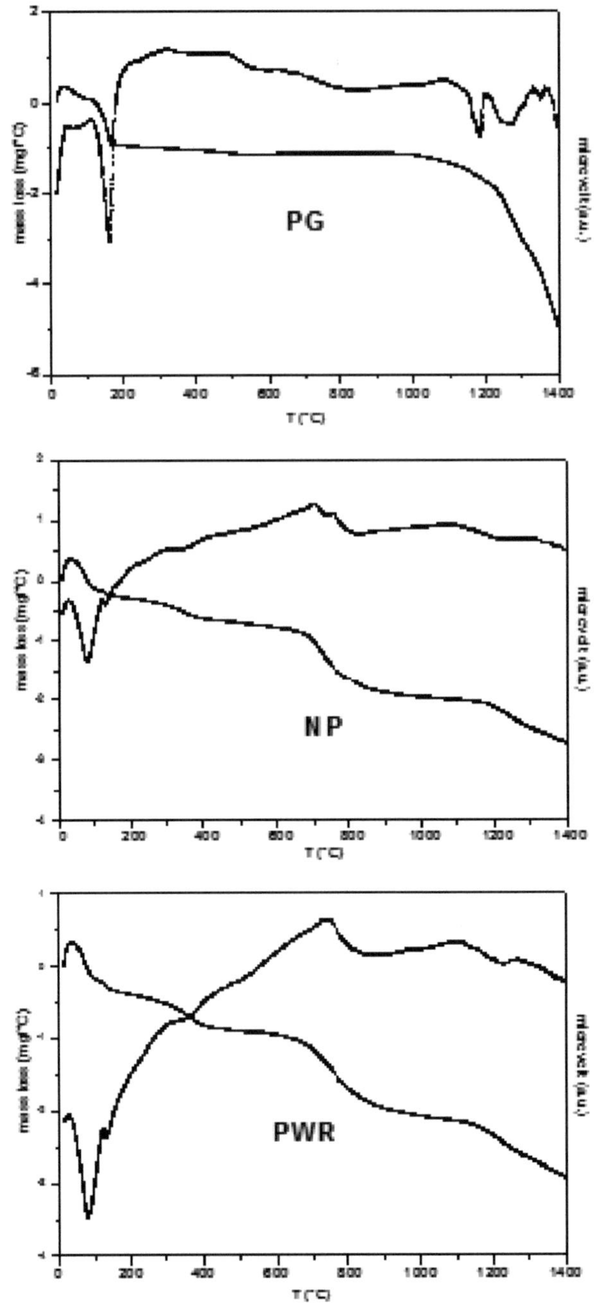

Figure 3.
TGA and (A') DTA features of phosphate waste rock. (PWR) natural phosphate, (NP) and phosphogypsum, (PG).

3.2 Removal studies

3.2.1 Effects of solid dosage

A series of runs were carried out by varying the used solid mass from 0.05 to 2 g in 200 mL of AR 88 solution (C_i of 20 mg L^{-1}). The removal efficiency (%) of natural phosphate, phosphogypsum, and phosphate waste rock on the removal of

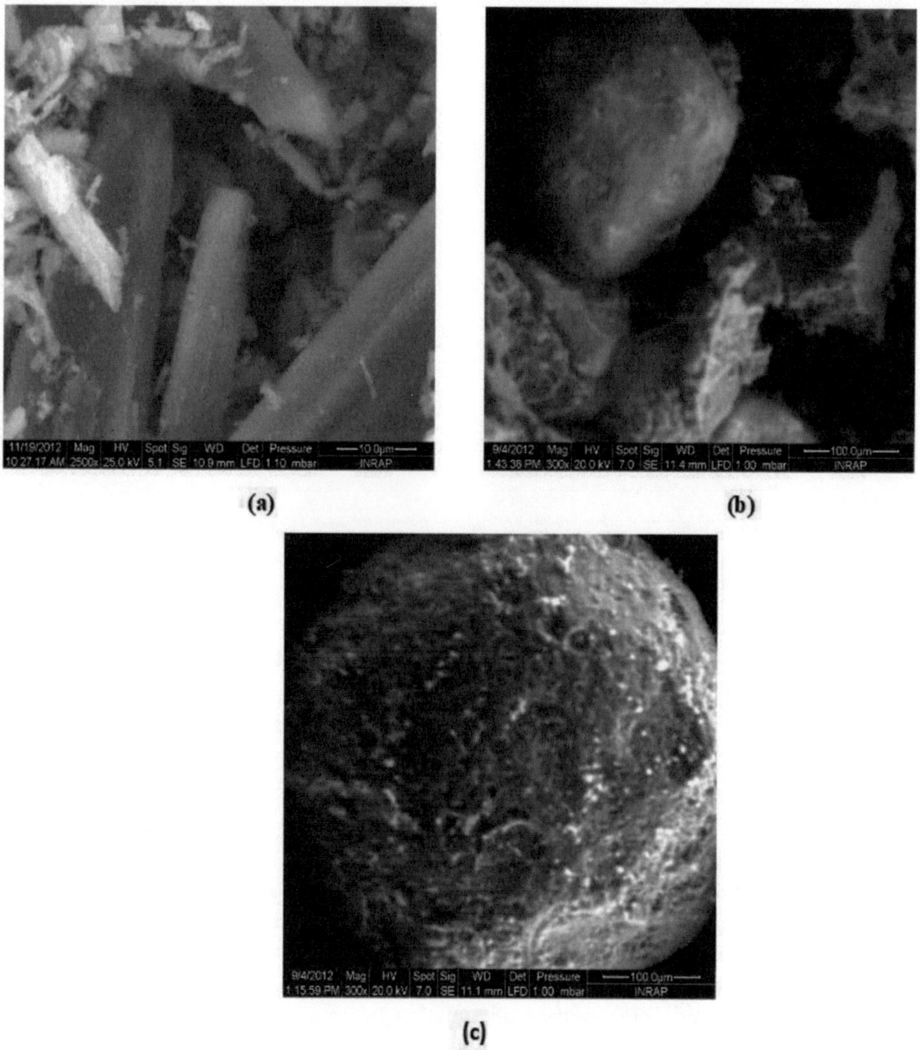

Figure 4.
SEM micrographs of (a) phosphogypsum (b), natural phosphate, and (c) phosphate waste rock.

Samples	S_{BET} (m^2/g)	T.P.V (cc/g)	A.P.D. (nm)
NP	11.35	0.027	9.58
PG	16.00	0.031	7.65
PWR	26.43	0.052	7.62

T.P.V = total pore volume; A.P.D. average pore diameter.

Table 2.
Micro textural properties of the different materials.

AR88 dye improves as the amount of added solid increases; this is due to the greater availability of active sites on the solid's surface (**Figure 5**) [51]. In particular, the phosphate waste rock on the removal of AR88 dye mate-rial exhibits a significant increase in both dyes removal efficiency, who reached 99% when the mass of used phosphate waste rock on the removal of AR88 dye is 1 g, However, the natural

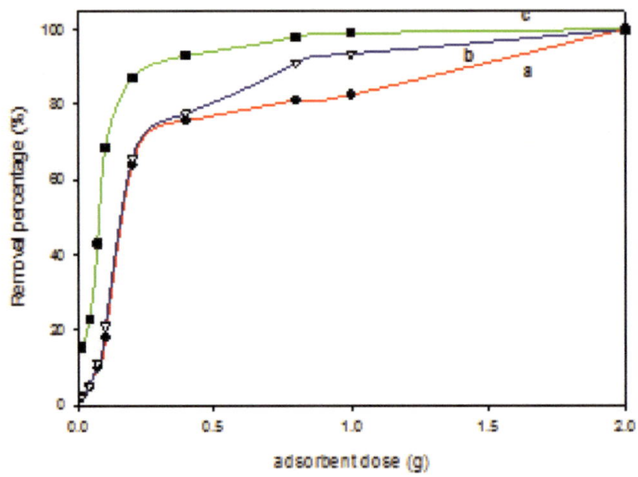

Figure 5.
Effect of dosage mass of (a) natural phosphate, (b) phosphogypsum, and (c) phosphate waste rock on the removal of AR88 dye.

phosphate and phosphogypsum materials exhibit similar removal efficiencies (99%) using a dose of 2 g due to their low removal capacities compared with phosphate waste rock. In general, increasing the removal dosage enhances the removal efficiency the dyes and attributed to the increase of the number of available removal sites.

3.2.2 Effect of pH

The work of the pH was studied at room temperature. The first step consists of a mixture of 1 g of the three adsorbents to 200 mL of AR 88 solution (Ci of 100 mg L^{-1}). In the second step the mixture was stirred for 240 minutes. The pH was altered between 2.5 and 11 using HCl (0.1 M) or NaOH (0.1 M) solutions.

The pH follow-ups the structure of the adsorbates and regulates the charge distribution of the samples [52], the zero-charge point (pzc) was determined first. The point of zero charge (pzc) of the three products NP, PG, and PWR is 6.89, 8.26, and 9.58, respectively. The results revealed that the surface particles are positively charged at pH values below pHpzc, while at pH values lower than pHpzc they become negatively charged.

When the pH diminishes the quantity of acid dye retained by the three adsorbents rises (**Figure 6**). This result can be explained by the dye structure and the protonation of the solid surface [53]. At acidic medium (pH = 3), it exists important electrostatic attractions between the ϕ-SO_3- groups of the positive surface charges of adsorbents and the dye molecules. Also, the OH^- anions additional are disposable and dispute with the anionic dye for available at higher pH values.

3.2.3 Kinetics of adsorption

The effect of contact time on the retention AR 88 respectively are showed in **Figure 7**. An amount of 1 g of the three samples was added to a volume of 200 mL of a 100 mg L^{-1} dye solution with stirring. The taken of the samples have been realized at several time during the reaction for 300 minutes.

The retention of the anionic dye is quick at a short time and the removal starts to get slow the equilibrium is reached after 240 minutes for the three materials.

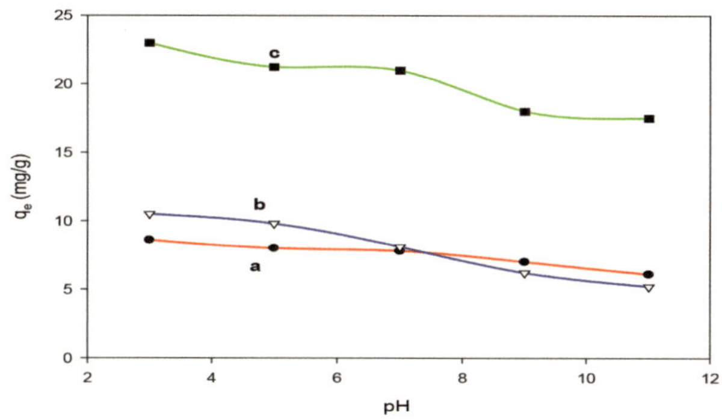

Figure 6.
Effect of initial pH on the removal of AR88 dye by (a) natural phosphate, (b) phosphogypsum, and (c) phosphate waste rock.

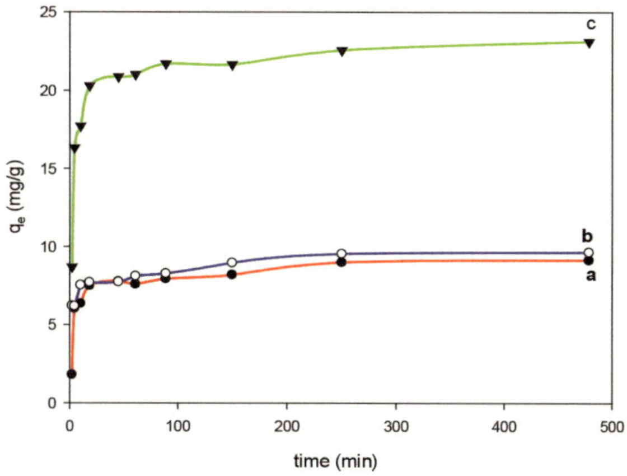

Figure 7.
Removal kinetics of AR88 on (a) natural phosphate, (b) phosphogypsum and (c) phosphate waste rock.

These results can be explained by the important number of existing sites for the acid dye during the removal processus.

Near to equilibrium, the number of sites is decreasing. What's more, the repulsive forces between the textile dye on the adsorbents and those in the solution are responsible of the to slow down of the speed adsorption [54, 55].

3.2.3.1 Pseudo-first-order kinetic model

This model describes the rate of change that occurs for the dye uptake [56, 57]. It is defined by Eq. (3):

$$\log\left(q_e - q_t\right) = \log q_e - \frac{k_1 t}{2,3} \tag{3}$$

where q_e and q_t are the removal capacities at equilibrium and time "t", respectively. k_1 is the first-order rate constant. The linear plot of log (q_e - q_t) vs. time "t" shows the applicability of this model for the retention of the acid dye. The parameter k_1 is illustrated in **Table 3**. This values of the three compounds are varied between 0.010 to 0.014 min^{-1}. The regression correlation coefficients R^2 are near to 0.900. While the experimental values of qe do not match the values predicted by this model.

Samples	Pseudo first order				Pseudo second order			
	K_1 (min^{-1})	q_m (mg.g^{-1})		R^2	K_2 (min^{-1}gmg^{-1})	q_m (mg.g^{-1})		R^2
		Calculated	Experimental			Calculated	Experimental	
NP	0.0107	3.343	8.75	0.903	0.006	8.849	8.75	0.998
PG	0.0109	3.536	9.26	0.904	0.080	9.216	9.26	0.997
PWR	0.0144	6.855	21.875	0.900	0.005	1.796	21.875	0.996

Table 3.
Constant rates of pseudo first order and pseudo second order for the removal of acid red 88 onto various samples.

3.2.3.2 Pseudo-second-order kinetic model

The pseudo-second-order kinetic model is illustrated in Eq. (4):

$$\frac{t}{q_t} = \frac{1}{k_2 q_e^2} + \frac{t}{q_e} \quad (4)$$

with qe and qt are the quantity of acid dye adsorbed at equilibrium and measured at time t (mg g^{-1}), respectively. k_2 (g mg^{-1} min^{-1}) is the pseudo-second-order rate constant. The various parameters of the pseudo-second-order model are given in **Table 3**. The k_2 values are varied between 0.005 and 0.008 g mg^{-1} min^{-1}. The calculated regression coefficient parameter (R^2) is near to 0.9964.

The values of the regression coefficients (R^2) near to 1 and the coincidence between the experimental results of q_e and the calculated values ones (**Table 3**) show that the retention process of anionic dye by the three compounds is characterized by the pseudo-second-order model and no than the pseudo-first–order kinetic type. Analog results have been observed for the adsorption of several textiles dyes [58–60].

3.2.4 Effect of initial concentrations

The impact of varying concentration of AR 88 on the adsorbed quantities onto PN, PG and PWR crudes of acid and basic dyes are showed in **Figure 8**. The adsorbed quantity of the acid textile dye hits a value near to 105 mg.g^{-1} when the C of AR 88 is 200 mg L^{-1} for the byproduct of a phosphate company's washing plant

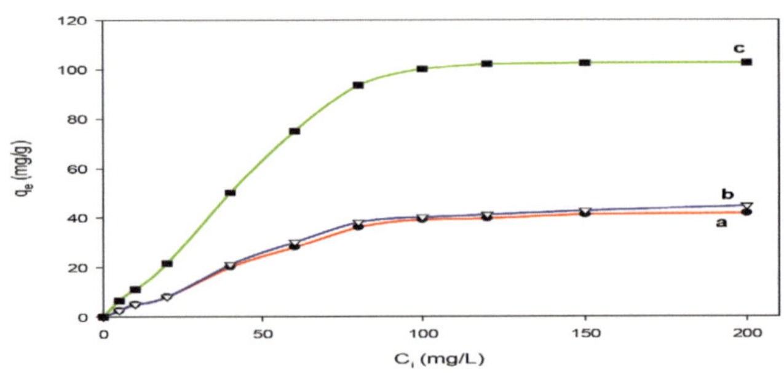

Figure 8.
Variation of removed amount of AR88 (q_e (mg/g)) as in function of initial concentration (C_i (mg/L)) using (a) natural phosphate, (b) phosphogypsum, and (c) phosphate waste rock at room temperature.

sample. For the Natural Phosphate and phosphorgypsum samples, the retained dye diminished to 40 and 43 mg g^{-1}, respectively.

At lower concentrations, the amount of the sites on the surface of the adsorbents compared to the number of dye molecules in the solution is important, and consequently, the acid dyes products interact with the samples. At higher concentrations, the AR 88 dye will be unavailable to contact surface sites filled [61].

3.2.5 Isotherms models

The analysis of the isotherm is an important step to optimize the design of the removal process [62]. Langmuir and Freundlich are often used to describe equilibrium isotherms. The Langmuir model is commonly applied to a complete homogeneous surface when the interaction between adsorbed molecules is negligible [63].

3.2.5.1 Langmuir isotherm model

The linear equation of the Langmuir model is expressed in Eq. (5):

$$\frac{C_e}{q_e} = \frac{1}{q_{\max}.K_L} + \frac{C_e}{q_{\max}} \tag{5}$$

where q_e and C_e are the removed amount of dye (mg/g) and the concentration (mg L^{-1}) in the solution at equilibrium, respectively. q_{max} is the maximum removed amount (mg/g), and K_L (L/mg) is the Langmuir constant. The linear plot of Ce/qe versus C_e was used to evaluate these constants.

In the case of the Langmuir model, the constants obtained for the removal of the AR 88 dye by the NP, PG, and PWR materials are presented in **Table 4**. The values of regression correlation coefficients (R2) are higher than 0.999. These values revelated that the retention of the acid dye by the three compounds from the phosphate industry is accurately described by the Langmuir model. The estimated maximum removal capacities (q_m) are 48.4, 49.0, and 123.4 mg/g for NP, PG, and PWR, respectively. Moreover, the K_L values range between 0.032 and 0.035 L/mg.

A dimensionless constant separation factor or equilibrium parameter "R_L", a characteristic of a Langmuir isotherm, is defined in Eq. (6):

$$R_L = \frac{1}{1 + K_L C_i} \tag{6}$$

with C_i is the initial concentration (mg L^{-1}), and KL is the Langmuir constant (L/mg). Parameters of R_L in the range zero and one ($0 < R_L < 1$) reveals that the

Isotherm model	Parameters	NP	PG	PWR
Langmuir	$q_{m\ exp}$ (mg g¯1)	48.40	49.00	123.45
	K_L (L mg^{-1})	0.032	0.033	0.035
	R^2	0.997	0.993	0.994
Freundlich	K_F (L mg^{-1}	0.729	0.671	1.459
	n	0.83	0.85	0.84
	R^2	0.978	0.976	0.978

Table 4.
Langmuir and Freundlich constants, for the removal of acid red 88 onto various samples.

retention is favorable; the adsorption is linear for RL = 1, unfavorable for R_L greater than 1, and irreversible when R_L is equal to 0. In our case, R_L values were determined to be between 0 and 1, indicating the favorable adsorption of the dye to all adsorbents [64].

3.2.5.2 Freundlich isotherm model

The removal of AR88 by the different materials was also fitted to the Freundlich model [65] with the linear equation in Eq. (7):

$$\text{Lnqe} = \text{LnKF} + \frac{1}{n}\text{LnCe} \tag{7}$$

with the adsorbed quantity (qe, $mg.g^{-1}$) is linearly joined to the concentration of the anionic dye at equilibrium (Ce), and K_F and 1/n are the Freundlich constants. K_F is a combined measure of both the retention capacity and affinity, and 1/n informs about the degree or intensity of the retention of anionic dye. The favorability of the retention is given by the magnitude of n, i.e., values of 1/n less than 1 ($0 < 1/n < 1$) [66].

The values of the determined Freundlich parameters are given in **Table 4**. The acid dye affinity measured by the coefficient (OF) is in the order, phosphate waste rock > natural phosphate > phosphogypsum. The values of the 1/n parameter are inferior to 1, revealing that the retention of acid dye is favorable in operators' conditions. The R^2 values determined by the Freundlich model are near to 0.971 inferior that those calculated by the Langmuir model revelation that the experimental data fit well to the Langmuir isotherm model.

The obtained results indicate that, during the retention of textile dye, the last product is transferred to energetically equivalent sites, with the acid dye molecules forming a monolayer on the outer surface of the used adsorbents.

3.2.6 Effect of adsorption temperature

Temperature is a crucial parameter that affects the removal process and enables the determination of thermodynamic parameters. The effect of temperature on the adsorbed amount of NP, PG and PWR were investigated. A series of experiments were performed while maintaining the concentration of AR 88 at 200 mg L^{-1}.

The removal of AR 88 decreased at equilibrium with an increase in temperature, indicating that the removal is an exothermic process [67]. The removal process in the case is an endothermic [67]. Changes in thermodynamic parameters, such as (ΔG°), (ΔH°), and (ΔS°), were calculated by the following equations [68, 69]:

$$\Delta G^\circ = \Delta H^\circ + T\Delta S^\circ \tag{8}$$

$$\Delta G^\circ{}_{ads} = -RTlnK_c \tag{9}$$

$$\ln K_c = \left(\frac{\Delta S^\circ}{R}\right) - \left(\frac{\Delta H^\circ}{R}\right)\frac{1}{T} \tag{10}$$

where K_c is the distribution coefficient of AR 88 removal from aqueous solution by waste materials, "T" is the absolute temperature, and R is the gas constant.

Table 5 summarizes the estimated thermodynamic parameters. The negative values of ΔG° at various temperatures indicate the spontaneous nature of the removal process. The negative values of H° confirm that the removal of dye using the various samples is an exothermic process [68].

Samples	H° (kJ mol^{-1})	S° (kJ mol^{-1})	G° (kJ mol^{-1})	R^2
NP	−13.082	−37.805	13.08	0.985
PG	−15.335	−48.223	−15.33	0.992
PWR	−83.133	−16.255	−83.13	0.997

Table 5.
Thermodynamic parameters for the retention of anionic dye on various compounds.

The negative ΔS° accompanying the removal of AR 88 indicates a less disordered system accompanied by a reduction in the randomness of the dye molecules at the solid–liquid interface.

The negative values of ΔG° indicated the spontaneous nature of the removal process. Similar data were reported for different used adsorbents. The ΔG° values are −15.33 kJ mol^{-1} (for PG) and – 13.08 kJ mol^{-1} (for NP) in the case of the removal of AR 88 dye. These values represent major physical adsorption [70]. On the other hand, the removal process with PWR for the retention of the AR 88 dye characterized by chemical adsorption, where the change in free energy (ΔG°) value is −83.13 kJ mol^{-1}. This process involves strong forces of attraction [71, 72]. The rise of the change in free energy (ΔG°) values with temperature could be attributed to a diminish in the molecular order during the removal process.

4. Regeneration data

The recycle of adsorbents is a very important and crucial factor to propose an efficient adsorbent. A good adsorbent supposes to have higher removal capacity as well as regeneration efficiency that will reduce the total cost of the wasted adsorbent [73].

Different methods of regeneration were reported in the literature, including two ways, washing of the used samples with different solutions such ethanol, acidic or basic ones to remove the adsorbed dyes as they are [73, 74], or to destroy the adsorbed dyes by thermal treatment at certain temperatures [75], this process will add additional costs to the process due the extra energy consumption. Another method was proposed by other researchers to destroy the adsorbed dyes on the surface of the solids via sulphate radical oxidation. This method was reported to be friendly to the environment since the solution could be used for many regeneration tests [26].

After the step of the retention of the anionic dye, the regeneration and consumption of the waste adsorbents indicates the valuable and the feasibility of their application. The **Figure 9**, show the adsorption efficiency is decreased, between 90% and 85%, regarding the byproduct of a phosphate washing sample for at least four try. Also, the other two samples: Natural Phosphate and phosphogypsum, reveals their retention percentages diminish up to 70% after three regeneration cycles. Overall, the retention efficiency is maintained at 60% for the seventh regeneration cycle.

The study of the adsorption isotherms can be used to study the removal systems. The design objective was to minimize the solid adsorbent for a specific volume of initial concentration.

The study of the adsorption isotherms can be used to study the removal systems. The aim of this study was to reduce the solid adsorbent for a specific volume of initial concentration.

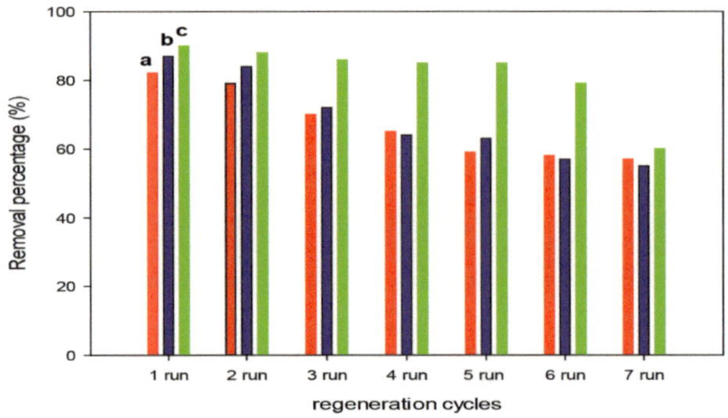

Figure 9.
The percentage removal of AR88 after different regeneration cycles, (a) natural phosphate, (b) phosphogypsum, and (c) phosphate waste rock.

Consider an effluent containing V (L) of the solution in contact with the colorant and let the dye concentration got reduced from C_0 to C_1 mg dye L^{-1} solution. For a quantity of adsorbent m (g), the solute loading changed from q_0 to q_1 (mg dye per g adsorbent). When fresh adsorbent is used, $q_0 = 0$, the mass balance for the methylene blue (MB) dye in the single-stage operation under equilibrium is presented in Eq. (11).

$$V(Co - Ce) = m(q_o\ q_e) = mqe \tag{11}$$

In the present case, the removal of AR 88 corresponded well with the Langmuir isotherm. Consequently, the Langmuir equation can be substituted in the Eq. (5), and the rearranged form is given in Eq. (12).

$$\frac{m}{V} = \frac{C_o - C_e}{q_e} = \frac{C_o - C_e}{\frac{q_m K_L C_e}{1 + K_L C_e}} \tag{12}$$

Figure 10 represent the plots derived from Eq. (12) to predict the amount of phosphogypsum and PWR required (g) to treat different effluent volumes of the

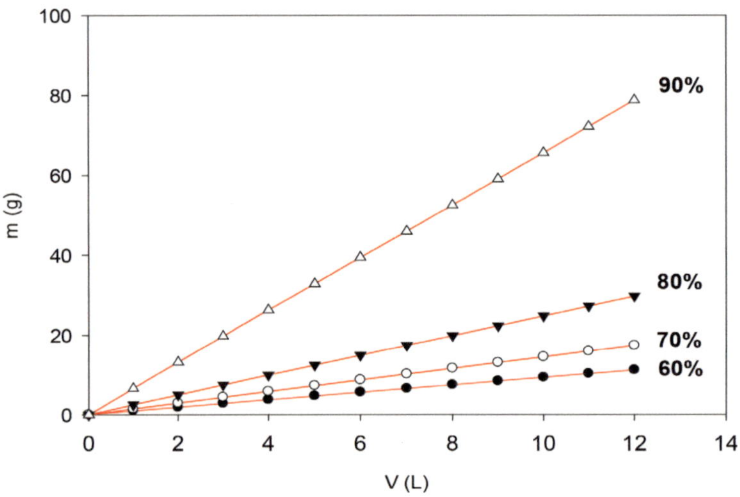

Figure 10.
Predicted mass (m) of PG waste to treat different volumes (V) of AR-88 solutions, at initial concnetration of 100 mg/L.

initial concentration of 100 mg L^{-1} for 60%, 70%, 80%, and 90% MB removal at different MB solution volumes from 1 to 12 L in 1 L increment. For a single design, the amount of phosphate wastes can be predicted in the range of 11 to 78.78 g for phosphogypsum and 4 to 24.97 g for phosphate waste rock materials. That is to say that the amount required for 90% removal of MB solution of the initial concentration of 100 mg L^{-1}, was about 78.78 g of PG, and 24.97 g of PWR solids, respectively. The lower mass's values for phosphate waste rock solid were associated to their higher efficiency to remove acid dyes compared to phosphogypsum waste. These data indicated that these materials could be useful as removal agents for anionic dye.

5. Conclusions

The use of byproducts from the phosphate industry could create opportunities for the treatment of water contaminated by textile dyes. This study shows that natural phosphate, phosphogypsum, and phosphate waste rock are indeed appropriate for anionic dye removal. However, phosphate waste rock (123.4 mg.g^{-1}) has a removal capacity that is higher than that of the natural phosphate (48.4 mg.g^{-1}) and phosphogypsum (49.0 mg.g^{-1}) materials.

The removal is dependent on the pH of the dye solution, with higher uptake of the dye at a lower pH in the case of the acid dye. The quantity retained increases when the pH increases for the basic dye.

The removal rate of the acid dye fits the pseudo-second-order model for all three materials. The Langmuir isotherm model more appropriately explains the experimental data while, it suggests the formation of a dye monolayer on the surface of the waste products. The Freundlich model isotherm explained appropriately the experimental data, and it suggests the heterogeneity of the surface and presumes that the adsorption takes place at sites with different adsorption energies.

n the case of acid dye, the thermodynamic parameters revealed that the removal process is spontaneous, exothermic, and occurred via chemisorption with the phosphate waste rock. However, physisorption is the proposed mechanism of removal with natural phosphate and phosphogypsum.

For the anionic dye, the regeneration of waster byproducts reveals that close to 80% of the retention dye was adsorbed after four cycles for the phosphate waste rock compound, and it was diminished to 60% after its consumption for seven cycles for all the adsorbents. A trials design was proposed, these experiences are based on the Langmuir model and the high adsorbed quantity. The sought mass of phosphate reuses to achieve a constant percentage of cationic Methylene Blue dye retention, could be readily necessary. The quantity values is linked to the phosphate reuse because their difference in their retention efficiency. Nevertheless, these reported data suggested that wastewater treatment is a potential application for the waste products of phosphate mining.

Author details

Khaled Boughzala[1,2]* and Mustapha Hidouri[3]

1 Unité de Recherche Analyses et Procédés Appliqués à l'environnement, Institut Supérieur des Sciences Appliquées et Technologie de Mahdia, Mahdia, Tunisia

2 Unit Electrochemistry, Materials and Environment, University of Kairouan, Kairouan, Tunisia

3 High Institute of Applied Sciences and Technology, Gabes, Tunisia

*Address all correspondence to: khaledboughzala@gmail.com

References

[1] K. Hunger, Industrial Dyes: Chemistry, Properties, Applications. Wiley-Verlag GmbH & Co, Weinheim, 2003

[2] O.M.L. Alharbi, A.A. Basheer, R.A. Khattab and I. Ali, Health and environment effects of persistent organic pollutants, J. Mol. Liq., 263 (2018) 442-453.

[3] S.J. Culp and F.A. Beland. Malachite Green: A Toxicological Review, Inter. J. Toxic., 15 (1996) 219-238.

[4] A.E. Ghaly, R. Ananthashankar, M. Alhattab, V.V. Ramakrishnan, Production, characterization and treatment of textile effluents: a critical review, J. Chem. Eng. Process. Technol., 5 (2014) 182, doi: 10.4172/2157-7048.1000182.

[5] T. Asano, A.D. Levine, Wastewater reclamation, recycling and reuse: past, present, and future, Water Sci. Technol., 33 (1996) 1–14.

[6] F.R. Rijsberman, Water scarcity: fact or fiction?, Agric. Water Manage., 80 (2006) 5–22.

[7] A. Rossner, S.A. Snyder and D.R.U. Knappe, Removal of emerging contaminants of concern by alternative adsorbents, Water Res., 43 (2009) 3787-3796.

[8] S. Venkata Mohan, P. Sailaja, M. Srimurali and J. Karthikeyan, Colour removal of monoazo acid dye from aqueous solution by adsorption and chemical coagulation, Environ. Eng. Policy, 1 (1999) 149-154

[9] G. Crini, Non-conventional low-cost adsorbents for dye removal: A review, Bioresource Technol. 97 (2006) 1061-1085.

[10] M. Muthukumar, D. Sargunamani and N. Selvakumar, Research article Abstract only Statistical analysis of the effect of aromatic, azo and sulphonic acid groups on decolouration of acid dye effluents using advanced oxidation processes, Dyes Pigments 65 (2005) 151-158.

[11] M.S. Khehra, H.S. Saini, D.K. Sharma, B.S. Chadha and S.S. Chimni, Biodegradation of azo dye C.I. Acid Red 88 by an anoxic–aerobic sequential bioreactor, Dyes Pigments 70 (2006) 1-7

[12] C. O'Neill, A. Lopez, S. Esteves, F.R. Hawkes, D.L. Hawkes and S. Wilcox., Azo-dye degradation in an anaerobic-aerobic treatment system operating on simulated textile effluent, Appl. Microbiol. Biot. 53 (2000) 249-254.

[13] H.D. Beyene, The potential of dyes removal from textile wastewater by using different treatment technology: a review, Int. J. Environ. Monit. Anal., 2014 (2014) 347–353.

[14] M.T. Yagub, T.K. Sen, S. Afroze, H. M. Ang, Dye and its removal from aqueous solution by adsorption: a review, Adv. Colloid Interface Sci., 209 (2014) 172–184.

[15] G.Z. Kyzas, J. Fu, K.A. Matis, The change from past to future for adsorbent materials in treatment of dyeing wastewaters, Materials, 6 (2013) 5131–5158.

[16] M.A.M. Salleh, D.K. Mahmoud, W. A.W.A. Karim, A. Idris, Cationic and anionic dye adsorption by agricultural solid wastes: a comprehensive review, Desalination, 280 (2011) 1–13.

[17] C.A.P. Almeida, A. dos Santos, S. Jaerger, N.A. Debacher, N.P. Hankins, Mineral waste from coal mining for removal of astrazon red dye from aqueous solutions, Desalination, 264 (2010) 181–187.

[18] M.A. Rauf, I. Shehadeh, A. Ahmed, A. Al-Zamly, Removal of methylene blue from aqueous solution by using gypsum as a low cost adsorbent, World Acad. Sci. Eng. Technol., 3 (2009) 540–545.

[19] Y.H. Wu, J.L. Cao, P. Yilihan, Y.P. Jin, Y.J. Wen, J.X. Zhou, Adsorption of anionic and cationic dyes from single and binary systems by industrial waste lead–zinc mine tailings, RSC Adv., 3 (2013) 10745–10753.

[20] S.K. Giri, N.N. Das, G.C. Pradhan, Magnetite powder and kaolinite derived from waste iron ore tailings for environmental applications, Powder Technol., 214 (2011) 513–518.

[21] J. Cooper, R. Lombardi, D. Boardman, C. Carliell-Marquet, The future distribution and production of global phosphate rock reserves, Resour. Conserv. Recycl., 57 (2011) 78–86.

[22] P. Becker, Phosphates and Phosphoric Acid, Raw Materials Technology, and Economics of the West Process, Marcel Dekker Inc., New York, 1989.

[23] N. Abbes, E. Bilal, L. Hermann, G. Steiner and N. Haneklaus, Thermal beneficiation of Sra Ouertane (Tunisia) low-grade phosphate rock, Minerals, 2020, 10, 937-950.

[24] N. Barka, A. Assabbane, A. Nounah, L. Laanab, Y. Aît Ichou, Removal of textile dyes from aqueous solutions by natural phosphate as a new adsorbent, Desalination, 235 (2009) 264–275.

[25] A. Achkoun, J. Naja, R. M'Hamdi, Elimination of cationic and anionic dyes by natural phosphate, J. Chem. Eng., 6 (2012) 721–725.

[26] F. Kooli, Y. Liu, M. Abboudi, H. Oudghiri-Hassani, S. Rakass, S.M. Ibrahim, F. Al-Wadaani, Waste bricks applied as removal agent of Basic Blue 41 from aqueous solutions: base treatment and their regeneration efficiency, Appl. Sci., 9 (2019) 1237

[27] A.Z.M. Abouzied, Physical and thermal treatment of phosphate ores-An overview. International J. Mineral Processing. 85(2008) 59–84.

[28] H.H. Lim, Beneficiation of apatite rock phosphates by calcination: Effects on chemical properties and fertilizer effectiveness. Australian J. Soil Res. 39 (2001) 397-402.

[29] P.M. Rutherford, M.J. Dudas, R.A. Samek, Environmental impacts of phosphogypsum, Sci. Total Environ., 149 (1994) 1–38.

[30] P.M. Rutherford, M.J. Dudas, R.A. Samek, Environmental impacts of phosphogypsum, Sci. Total Environ., 149 (1994) 1–38.

[31] H. Tayibi, M. Choura , F.A. López, F.J. Alguacil and A. López-Delgado, Environmental impact and management of phosphogypsum, J. Environm. Manag., 90(2009) 2377-2386.

[32] Z. Graba, S. Hamoudi, D. Bekka, N. Bezzi an R. Roukherroub, Influence of adsorption parameters of basic red dye 46 by the rough and treated Algerian natural phospahtes, J. Industrial Eng. Chem., 25 (2015) 229-238.

[33] A. Aklil, M. Mouflih, and S. Sebti, Removal of heavy metal ions from water by using calcined phosphate as a new adsorbent, J. Hazard. Mater., 112 (2004) 183-119.

[34] N.S. Labidi and N.E. Kacemi. Adsorption mechanism of malachite green onto activated phosphate rock: a kinetics and theoretical study, Bulletin of Environmental Studies, 1(2016) 69-74.

[35] K. Boughzala, F. Kooli, N. Meksi, A. Bechrifa and K. Bouzouita. Waste

products from the phosphate industry as efficient removal of Acid Red 88 dye from aqueous solution: their regeneration uses and batch design adsorber, Desal. Water treat., 202 (2020) 410-419.

[36] M.A. Rauf, I. Shehadeh, A. Ahmed, and A. Al-Zamly, Removal of Methylene Blue from aqueous solution by using gypsum as a low cost adsorbent. World Acad. Sci. Eng. Technol., 3 (2009), 540-545.

[37] Y. Wu, J. Cao, P. Yilihan, Y. Jin, Y. Wen and J. Zhou, Adsorption of anionic and cationic dyes from single and binary systems by industrial waste lead-zinc mine tailings. RSC Advances, 3 (2013) 10745-10753.

[38] N. Barka, A. Assabbane, A. Nounah, L. Laanab and Y. AîtIchou, Removal of textile dyes from aqueous solutions by natural phosphate as a new adsorbent, Desalination, 235 (2009) 264-275.

[39] Z. Elouear, J. Bouzid, N. Boujelben, M. Feki, F. Jamoussi and A. Montiel, Heavy metal removal from aqueous solutions by activated phosphate rock, J. Hazard. Mater., 156 (2008) 412-420.

[40] E. Keleş, A.K. Özer and S.Yörük, Removal of Pb2+ from aqueous solutions by phosphate rock (low-grade). Desalination, 253 (2010) 124-128.

[41] N.S. Labidi and N.E. Kacemi, Equilibrium modelling and kinetic studies on the adsorption of basic dye by natural and activated algerian phosphate rock, Environmental Research International, 2 (2016) 1-6.

[42] I. Bouatba, L. Bilali, M. Benchanaa and M. El-Hammioui, Decadmiation of natural phosphates by heat treatment and hydrochloric acid. Asian J. Chem. 28, (2016), 819-824.

[43] A. Mizane and A. Louhi, Calcination effects on sulfuric dissolution of phosphate extracted from djebel onk mine (Algeria), Asian J. Chem. 20, (2008), 711-717

[44] N. Gmati, K. Boughzala, M. Abdellaoui, and K. Bouzouita, Mechanochemical synthesis of strontium britholites: Reaction mechanism. Comptes Rendus Chimie, 14 (2011) 896-903.

[45] A.A. Hanna, A.I.M. Akarish, S.M. Ahmed, Phosphogypsum: part I: mineralogical, thermogravimetric, chemical and infrared characterization, J. Mater. Sci. Technol., 15 (1999) 431–434.

[46] J.P. Lafon, E. Champion, D. Bernache-Assollant, Processing of AB-type carbonated hydroxyapatite Ca10–x (PO4)6–x(CO3)x(OH)2– x–2y(CO3)y ceramics with controlled composition, J. Eur. Ceram. Soc., 28 (2008) 139–147.

[47] Z. Graba, S. Hamoudi, D. Bekka, N. Bezzi, R. Boukherroub, Influence of adsorption parameters of basic red dye 46 by the rough and treated Algerian natural phosphates, J. Ind. Eng. Chem., 25 (2015) 229–238.

[48] N. Bezzi, D. Merabet, N. Benabdeslem, H. Arkoub, Caractérisation physico-chimique du minerai de phosphate de Bled el Hadba - Tebessa, Ann. Chim. Sci. Mater., 26 (2001) 5–23.

[49] L. S. Sebbahi, M. Lemine Ould Chameikh, F. Sahban, J. Aride, Benarafa, L. Belkbir, Thermal behaviour of Moroccan phosphogypsum, Thermochim. Acta, 302 (1997) 69–75.

[50] H. Bouyarmane, S. Saoiabi, A. Laghzizil, A. Saoiabi, A. Rami, El-Karbane, Natural phosphate and its derivative porous hydroxyapatite for the removal of toxic organic chemicals, Desal. Water Treat., 52 (2014) 7265–7269.

[51] F. Kooli, L. Yan, R. Al-Faze, A. Al-Sehimi, Removal enhancement of

basic blue 41 by brick waste from an aqueous solution, Arabian J. Chem., 8 (2015) 333–342.

[52] A. Chadlia, M. Mohamed Farouk, Removal of basic blue 41 from aqueous solution by carboxymethylated Posidonia oceanica, J. Appl. Polym. Sci., 103 (2007) 1215–1225.

[53] J.X. Lin, L. Wang, Adsorption of dyes using magnesium hydroxide-modified diatomite, Desal. Water Treat., 8 (2009) 263–271.

[54] M.J. Martin, A. Artola, M.D. Balaguer, M. Rigola, Activated carbons developed from surplus sewage sludge for the removal of dyes from dilute aqueous solutions, Chem. Eng. J., 94 (2003) 231–239.

[55] S. Rakass , H. Oudghiri Hassani, M. Abboudi, F. Kooli, A. Mohmoud, A. Aljuhani and F. Al Wadaani, Molybdenum Trioxide: Efficient Nanosorbent for Removal of Methylene Blue Dye from Aqueous Solutions, Molecules, 23 (2018) 2295- 2308.

[56] F. Kooli, L. Yan, R. Al-Faze and A. Al Suhaimi, Effect of acid activation of Saudi local clay mineral on removal properties of basic blue 41 from an aqueous solution, Applied Clay Sci., 116-117 (2015) 23-30.

[57] S. Lagergren, About the theory of so-called adsorption of soluble substances, Kungliga Svenska Vetenskapsakademiens Handlingar, 24 (1898) 1–39.

[58] H.El Boujaady, A. El Rhilassi, M.Bennani-Ziatni, R.El Hamri, A.Taitai and J.L.Lacout,Removal of a textile dye by adsorption on synthetic calcium phosphates, Desalination, 275 (2011) 10-16.

[59] Y.S. Ho, T.H. Chiang, Y.M. Hsueh, Removal of basic dye from aqueous solution using tree fern as a biosorbent, Process Biochem., 40 (2005) 119–124.

[60] M. Doğan, M. Alkan, Adsorption kinetics of methyl violet onto perlite, Chemosphere, 50 (2003) 517–528.

[61] M. Al-Ghouti, M.A.M. Khraisheh, M.N.M. Ahmad, S. Allen, Thermodynamic behaviour and the effect of temperature on the removal of dyes from aqueous solution using modified diatomite: a kinetic study, J. Colloid Interface Sci., 287 (2005) 6–13.

[62] V. Vimonses, B. Jin, C.W.K.Chow and C. Saint, Enhancing removal efficiency of anionic dye by combination and calcination of clay materials and calcium hydroxide, J. Hazrd. Mater.. 171 (2009) 941-947.

[63] K.H. Foo, B.H. Hameed, Insights into the modeling of adsorption isotherm systems, Chem. Eng. J., 156 (2010) 2–10.

[64] I. Langmuir, The adsorption of gases on plane surfaces of glass, mica and platinum, J. Am. Chem. Soc., 40 (1918) 1361–1403.

[65] A. Bera, T. Kumar, K. Ojha, A. Mandal, Adsorption of surfactants on sand surface in enhanced oil recovery: Isotherms, kinetics and thermodynamic studies.Appl. Surf. Sci.2013, 284, 87-99.

[66] O. Hamdaoui,; E. Naffrechoux, Modeling of adsorption isotherms of phenol and chlorophenols onto granular activated carbon. Part I. Two-parameter models and equations allowing determination of thermodynamic parameters. J. Hazard. Mater. 2007, 147, 381-394.

[67] G. Rytwo, E. Ruiz-Hitzky, Enthalpies of adsorption of methylene blue and crystal violet to montmorillonite, J. Therm. Anal. Calorim., 71 (2002) 751–759.

[68] C.H. Wu, Adsorption of reactive dye onto carbon nanotubes: equilibrium,

kinetics and thermodynamics. J. Hazard. Mater.. 144 (2007) 100.

[69] A. Ramesh, D.J. Lee, and J.W. Wong, Thermodynamic parameters for adsorption equilibrium of heavy metals and dyes from wastewater with low cost adsorbents. J. Colloid Interface Sci. 291 (2005),588-592.

[70] W.J. Weber Jr, P.M. McGinley, L.E. Katz, Sorption phenomena in subsurface systems: concepts, models and effects on contaminant fate and transport, Water Res., 25 (1991) 499–528.65.

[71] M.A. Ferro-Garcia, J. Rivera-Utrilla, I. Bautista-Toledo, A.C. Moreno-Castilla, Adsorption of humic substances on activated carbon from aqueous solutions and their effect on the removal of Cr(III) ions, Langmuir, 14 (1998) 1880–1886.

[72] S.K. Ismadji, Bhatia, Adsorption of flavour esters on granular activated carbon, Can. J. Chem. Eng., 78 (2000) 892-901.

[73] R. Lafi, I. Montasser and A. Hafiane, Adsorption of congo red dye from aqueous solutions by prepared activated carbon with oxygen-containing functional groups and its regeneration, Adsorption Sci. Technol., 37(2019) 160-181.

[74] T.P.K. Kulasooriya, N. Priyantha and A.N. Navaratne, Removal of textile dyes from industrial effluents using burnt brick pieces: adsorption isotherms, kinetics and desorption. S.N Appl. Sci. 2 (2020), 1789 https://doi.org/10.1007/s42452-020-03533-0.

[75] F. Kooli, Y. Liu, M. Abouddi, S. Rakass, H. Oudgiri Hassani, S.M. Ibrahim, R. Al-Faze, Application of organo-magadiites for the removal of eosin dye from aqueous solutions: thermal treatment and regeneration, Molecules, 23 (2018) 2280.